D0090512

Acclaim for *ORIGINS*

"This is a terrific book on a fascinating and largely unexplored subject—the mysteries of prenatal development. It is lucid, scientifically accurate, and clearly and gracefully written. Combining good science and a personal perspective is rare, especially in writing about children and motherhood, but Annie Murphy Paul has accomplished it beautifully."

—Alison Gopnik, author of *The Scientist in the Crib* and *The Philosophical Baby*

"Can what we experience in the womb affect us for the rest of our lives? In a word: yes. As Annie Murphy Paul shows in this fascinating exploration of a new line of research, the fetus not only grows and develops in utero, it actively prepares for life in the world outside, reading signals the mother's body is sending about whether there will be plenty or want, hardship or happiness, and fashioning itself accordingly. The implications—for policy, for prenatal care, for parenting—are endlessly important.

—Liza Mundy, author of *Everything Conceivable: How Assisted Reproduction Is Changing Our World*

"That the behavior—even the emotions—of a pregnant woman can profoundly change her developing child is a remarkable idea. In this brilliant book, Annie Murphy Paul shows us that groundbreaking research on fetal origins is not a cause for fear or anxiety, but for wonder and even hope."

—Ethan Watters, author of *Urban Tribes* and *Crazy Like Us*

"Annie Murphy Paul's *Origins* delights from the first sentence onward. Engaging and fresh, it answers a host of compelling questions about what is really happening between mother and child as the outside world makes its way inside in those crucial nine months of fetal development—like why you may not want to drink from plastic bottles, and what happens when you're stressed, and what about the air we breathe. While *Origins* deserves a place on every expectant parent's bookshelf, it should be of deep interest to anyone who has ever spent time in a womb."

—Sue Halpern, author of *Can't Remember What I Forgot* and *Migrations to Solitude*

"A trek through the new frontier of fetal origins, with a smart, savvy, motivated guide—Annie Murphy Paul, pregnant with her second child and driven to figure out what's going on in there. She lucidly describes what scientists are learning about the life-long impact of those first nine months, from the mother's diet to her stress level, from the BPA in plastic to the call and response of the 'fetal-maternal dance.' Read it with pleasure, and brace yourself for some surprises."

—Robin Marantz Henig, author of *Pandora's Baby: How the First Test Tube Babies Sparked the Reproductive Revolution*

"One of the most influential environments on Earth lies within women's bodies, the still mysterious world of the womb where all of us spend the beginning months of our lives. In her fascinating book, Annie Murphy Paul explores this strange and wonderful first home, both as a science journalist investigating the critical first steps in human development and as an expectant mother thinking about how a child grows ready for the world outside. The combination and the lessons contained in both journeys make *Origins* an irresistible—and important—way to better understand ourselves."

—Deborah Blum, author of *Sex on the Brain* and *The Poisoner's Handbook* *f*P

Also by Annie Murphy Paul

The Cult of Personality:
How Personality Tests Are Leading Us to Miseducate
Our Children, Mismanage Our Companies,
and Misunderstand Ourselves

ORIGINS

HOW THE NINE MONTHS
BEFORE BIRTH SHAPE
THE REST OF OUR LIVES

ANNIE MURPHY PAUL

FREE PRESS

New York   London   Toronto   Sydney

FREE PRESS
A Division of Simon & Schuster, Inc.
1230 Avenue of the Americas
New York, NY 10020

First Free Press hardcover edition September 2010

FREE PRESS and colophon are trademarks of Simon & Schuster, Inc.

For information about special discounts for bulk purchases,
please contact Simon & Schuster Special Sales at 1-866-506-1949
or business@simonandschuster.com.

The Simon & Schuster Speakers Bureau can bring authors to your live event.
For more information or to book an event, contact the Simon & Schuster Speakers Bureau
at 1-866-248-3049 or visit our website at www.simonspeakers.com.

Manufactured in the United States of America

1   3   5   7   9   10   8   6   4   2

Library of Congress Control No.: 2010015249

ISBN 978-0-7432-9662-5
ISBN 978-1-4391-7184-4 (ebook)

NOTE TO READERS

This publication contains the opinions and ideas of its author. It is intended to provide
helpful and informative material on the subjects addressed in the publication. It is sold
with the understanding that the author and publisher are not engaged in rendering medical, health, or any other kind of personal professional services in the book. The reader
should consult his or her medical, health, or other competent professional before adopting
any of the suggestions in this book or drawing inferences from it.

The author and publisher specifically disclaim all responsibility for any liability, loss,
or risk, personal or otherwise, which is incurred as a consequence, directly or indirectly, of
the use and application of any of the contents of this book.

To my children

CONTENTS

ORIGINS

ONE MONTH

If you're going to ponder the mysteries of our origins—who we are, and how we got this way—you could pick a worse spot than Tot Lot One Hundred Five. A small playground and neighborhood gathering place, it's located near 105th Street in Riverside Park, a green ribbon that runs along the western edge of Manhattan, far from Central Park's crowded zoo and carousel. On this sunny July morning, it's a child's garden of elemental delights: sand, mud, and water that burbles from a great stone turtle like some primeval spring. I'm leaning back on the lot's wrought-iron fence, beholding the variety of human nature and human physiognomy on display, average age two and a half: there are the big, lumbering kids and the delicate fine-boned ones, the exuberant squealers and the wide-eyed watchers, the children running in crazed circles around the jungle gym and the ones reclining regally as pashas in their padded strollers.

My eyes alight on my own son, three-year-old Teddy, who is studiously constructing a many-turreted fort in the sandbox. Looking at his sturdy frame and his intent expression, I find myself musing once more on a familiar question. It came to me for the first time in the hospital, when I sat for hours next to my newborn's bassinet, watching him make whimsical, arched-brow faces in his sleep like a tiny mime. It reappeared a couple of years later when my toddler son began speaking, a development as surprising and fantastical as Dr. Doolittle's talking animals. Now here it is again, the ever-renewing riddle of parenthood: *What makes you the way you are?*

Other parents seem to know the answer. For one confident camp, it's genes. "Teddy is serious but a little dreamy sometimes, like John," says a friend of my husband, a classic absentminded professor. "Teddy got your writerly sensitivity," says one of my girlfriends. "And your stubbornness," adds another. It's as if his personality traits were lottery numbers drawn at conception, numbered Ping-Pong balls already settled into their slots. For another camp, it's all about nurture: the stimulating mobiles and the educational toys, the organic vegetables and the judiciously applied time-outs—these things, they say, make children who they are. Gazing around the playground at the parents clustered on park benches and perched on the edge of the sandbox, I imagine the two sides lining up for a rumble, getting ready to duke it out: "Genes!" "No, environment!" "Nature!" "Nurture!"

But lately I've begun to wonder about another source of influence, one that incorporates both nature *and* nurture: the conditions our children experienced while still in the womb. When I was pregnant with Teddy I felt an awareness of his particular presence, a sense that his individual development was already well under way. Now, when some new facet of his mental or emotional or physical self is revealed, I find myself pondering the possibility that it had its origins in utero. What if children—what if all of us—owe our constellation of traits not only to the genes we inherit at conception, and not only to the world we encounter after birth, but also to the nine eventful months in between?

Of course, no woman who is pregnant today can escape hearing the message that what she does affects her fetus. She hears it while at her doctor's appointments and while listening to the car radio going home, sees it in the pregnancy guidebook she reads in bed at night and in the newspaper that lands on her step in the morning. If she should happen to miss one of these urgent bulletins, rest assured that a friend will send it to her in an email or bring it up over lunch, or that her mother will call her up and tell her, "Did you hear they're saying that pregnant women shouldn't . . . ?"

Always, it seems, the influence wielded by a pregnant woman is of a negative kind; always she is one slipup away from harming her fetus. Today's pregnant woman could be forgiven for feeling that there's a vast conspiracy afoot, bent on controlling her every action, stripping her of every pleasure, and inducing guilt at every turn. At least as strong as the urge to freak out is the impulse to dismiss it all as bunk, as cynical scare tactics or paranoid hysteria that should sensibly be tuned out. During my pregnancy with Teddy, all these reactions jostled for room in my head. As my belly grew larger I watched, dismayed, as my world grew smaller and my choices fewer, constricted by anxiety and confusion.

And then: a very different perspective on pregnancy came to my attention.

When I'm not taking Teddy to the playground, I work as a science writer for newspapers and magazines. My job is to trawl the murky depths of the academic journals, looking for something shiny and new—a sparkling idea that catches my eye in the gloom. Starting a few years ago, I began noticing a dazzling array of findings clustered around the prenatal period. These discoveries were generating considerable enthusiasm among scientists, even as they overturned settled beliefs and assumptions about when and how human qualities emerge—our health, our intelligence, our temperaments. This research, I learned, comes from a burgeoning field known as "fetal origins," and its message is worlds away from the reprimands I had come to expect from popular books and articles about pregnancy.

Here there is a palpable sense of excitement, of horizons opening wide instead of clamping down tight. Here there is a necessary acknowledgment that things can go wrong during gestation—but also a dawning sense that intrauterine conditions make a lot of things go *right*, that the prenatal period is where many of the springs of health and strength and well-being are found. And here there is a recognition that there is no generically ideal pregnancy to aspire to (and, inevitably, to fall short of): there is instead a highly personal and particular shaping of the fetus for the specific world into which it will be welcomed.

The evidence in support of this new view of the prenatal period is arriving from a number of disciplines and a variety of types of investigations. There are animal experiments, in which variables can be tightly controlled and manipulated. There is epidemiological research, in which patterns emerge from the study of very large groups of people. There are studies based on so-called experiments of nature: real-world events that create fortuitous circumstances for investigation. There are economic analyses, produced by a growing number of economists who have turned their attention to the costs and benefits generated by prenatal experience. There is epigenetic research, an exciting new approach that examines how the behavior of genes is altered by the environment—modifications that are made without changing DNA, and that occur with special frequency in the prenatal period. And there is groundbreaking fetal research, carried out in the laboratory with the cooperation of pregnant women.

Such studies are turning pregnancy into something it has never been before: a scientific frontier. Obstetrics was once a sleepy medical specialty, and research on pregnancy a scientific backwater. Now the nine months of gestation are the focus of intense interest and excitement, the subject of an exploding number of journal articles, books, and conferences. All this activity is leading us to a striking new conception of the fetus, the pregnant woman, and the relationship between them. The fetus, we now know, is not an inert being—"the larval stage of human development," in the wry words of one fetal origins researcher—but an active and dynamic creature, responding and even adapting to conditions inside and outside its mother's body as it readies itself for life in the particular world it will soon enter. The pregnant woman is neither a passive incubator nor a source of always-imminent harm to her fetus, but a powerful and often positive influence on her child even before it's born. And pregnancy is not a nine-month wait for the big event of birth, but a crucial period unto itself—"a staging period for well-being and disease in later life," as one scientist puts it.

The notion of prenatal influences may conjure up frivolous attempts to enrich the fetus, like playing Mozart through headphones placed on a pregnant belly. In reality, the nine-month-long process of shaping and molding that goes on in the womb is far more visceral and consequential than that. Much of what a pregnant woman encounters in her daily life—the air she breathes, the food and drink she consumes, the emotions she feels, the chemicals she's exposed to—are shared in some fashion with her fetus. They make up a mix of influences as individual and idiosyncratic as the woman herself. The fetus incorporates these offerings into its own body, makes them part of its flesh and blood. And, often, it does something more: it treats these maternal contributions as *information*, as biological postcards from the world outside. What a fetus is absorbing in utero is not Mozart's *Magic Flute*, but the answers to questions much more critical to its survival: Will it be born into a world of abundance, or scarcity? Will it be safe and protected, or will it face constant dangers and threats? Will it live a long, fruitful life, or a short, harried one? The pregnant woman's diet and stress level, in particular, provide important clues to prevailing conditions, a finger lifted to the wind. The resulting tuning and tweaking of the fetus's brain and other organs are part of what give humans their impressive flexibility, their ability to thrive in environments as varied as the snow-swept tundra, the golden-grassed savanna—and the limestone canyons of Manhattan's Upper West Side.

This conception of pregnancy is new, new enough that it is still striving for acceptance in some quarters. But it is also very old. Most peoples, in most times and places, have believed that events and experiences before birth can shape the individual for good or ill. Many of these notions were highly fanciful: the ancient Greeks' belief that looking at statues and other works of art during pregnancy would lead to the birth of a beautiful child, for example, or the belief among eighteenth-century Britons that cravings experienced by a pregnant woman could leave a permanent mark on her offspring. But such convictions emerged from a widely shared understanding of the relationship between a pregnant woman and her fetus as intimate and reciprocal. There may be only one culture, in fact, in which this idea was roundly rejected: the scientific and medical culture of the modern West. For much of the twentieth century, many scientists and doctors in this part of the world held fast to a most unusual belief: that the human fetus is impervious to external influence, unfolding its developmental program with automatic efficiency and minimal outside interference.

This historically anomalous belief may have arisen out of a desire to separate the practice of medicine from the superstitions of midwifery. It was likely encouraged by the powerful metaphors of the industrial age: the fetus was assembled like a widget in a factory, these experts imagined, or constructed like a building according to a blueprint. And it was indirectly perpetuated by the great argument of this era, the century's intellectual main event: the nature versus nurture debate. Both sides in this epic matchup gave short shrift to the prenatal period. Proponents of nurture directed our attention to the environment of early childhood—but neglected to consider the crucial environment of the womb. Advocates of nature stressed the determinative power of genes—but they knew little about epigenetics, the modification of gene expression that operates most consequentially in utero. The resulting general neglect of the prenatal period led to some specific, and mistaken, beliefs—beliefs that are now being vigorously challenged by the science of fetal origins.

For example: until surprisingly recently, many doctors and scientists were convinced that the fetus was a "perfect parasite," skimming the nutrients it needed from its mother's body, unaffected by the quantity or kind of food she consumed. (A friend of mine who had a baby just a few years ago was told by her obstetrician that she could eat only lettuce for her entire pregnancy and her fetus wouldn't know the difference.) The fetal origins literature tells a very different story: that the fetus is in fact exquisitely sensitive to its mother's diet.

Another article of faith called into question by the new findings is the assumption that major illnesses—from heart disease to diabetes to cancer—are caused by a combination of bad genes and bad lifestyle (too much salty, fatty food, too little exercise) in adulthood. In fact there is a third risk factor of which we've taken too little account: the individual's experiences in the womb. Fetal origins research suggests that the lifestyle that influences the development of disease is often not only the one we follow as adults, but the one our mothers practiced when they were pregnant with us as well.

A final belief discredited by the new research is the notion that the fetus is safely sealed away in the womb, protected from all manner of pollutants and poisons by the ever-vigilant placenta. In fact, we're learning, the fetus inhabits the same world as adults— the world of alcohol and cigarettes, of polluted air and water, of industrial chemicals untested for their safety. The fetus's small size and immature state of development, as well as the permeability of the defense systems deployed around it by its mother, mean that individuals are more vulnerable to environmental toxins during the prenatal period than at any other time in their lives.

These once-common attitudes were far from harmless, of merely theoretical interest. They led to two of the greatest medical disasters in history: the tragedies caused by thalidomide and diethylstilbestrol (DES). These were drugs given to pregnant women in the belief that the fetus would be unaffected, when in fact many offspring went on to develop severe malformations or aggressive cancers as a result of their exposure. Nor have these attitudes completely disappeared. They linger on, for example, in the sluggish response of health authorities to the current threat posed to fetuses by the class of chemicals known as endocrine disruptors, found in plastics and other commonly used products.

Protecting fetuses from such dangers will be one crucially important outcome of fetal origins research. Even more significant, this field is producing knowledge that will allow us to actively promote the health, intelligence, and well-being of the next generation, and to better understand the origins of our own traits and characteristics. But this larger project has so far been lost in the sensational and scolding coverage of fetal researchers' findings. Their discoveries have been cast as one long ringing alarm bell, one long line of doctors in white lab coats, shaking their fingers at pregnant women: *No, Don't, Stop!* A host of exciting, thought-provoking explorations are reduced to a petty list of rules, censoriously enforced. We should recognize what's really happening here: a profound idea about human nature is being reclaimed, resurrected by science. For many centuries people have believed that there is continuity between the individual in utero and the individual in the world; now there is solid evidence that this ancient belief is correct, albeit in a far more complex and nuanced way than our ancestors ever imagined.

For me, this development has special significance: two weeks ago, I learned that I was pregnant again. During my first pregnancy, I had nothing but questions. This time, I'm determined to find some answers. Just how is the fetus shaped by a woman's behavior during pregnancy? How is it affected by her diet, her stress level, her emotional state, her exposure to chemicals? How can she minimize harm and maximize benefit? And what does the emerging science of fetal origins mean for us as individuals, as parents, and as members of society? In my investigations, I'll use all the tools at my disposal as a science writer—delving into the research literature, interviewing scientists, observing them at work.

But science can't tell us everything we need to know about this new perspective; there's always a gap where the hard evidence of the laboratory meets the soft flesh of our bodies. So I'll also embark on this exploration as a pregnant woman, someone who is living what she's learning about. I'll bring back discoveries from the cutting edge of fetal origins research and apply them in my own life; I'll be my own natural experiment.

Lastly, I'll examine changing notions of pregnancy as a student of culture and history, a denizen of both the sidewalk and the library. In this I'll be following the lead of the English poet and philosopher Samuel Taylor Coleridge, who in 1802 immersed himself in a book by physician Thomas Browne. "Surely we are all out of the computation of our age, and every man is some months elder than he bethinks him," Browne wrote, "for we live, move, have a being and are subject to the actions of the elements, and malice of diseases, in that other world, the truest microcosm, the womb of our mother."

So struck was Coleridge by this passage that he added his own enthusiastic note in the margin. "Yes! The history of man for the nine months preceding his birth would probably be far more interesting and contain events of greater moment, than all the three score and ten years that follow it," he scribbled. Coleridge's speculation is now being put to the test; the "history of man for the nine months preceding his birth" is at long last being written, and one small strand of it is being inscribed in me.

Lost in these thoughts, I feel a sudden tug on my hand. It's Teddy, his battlements at last completed. The sun is high in the sky; it's time to head home for lunch and a nap. As we swing the gate closed behind us, I take a last look over my shoulder at the parents and children in the park, all deeply engrossed in their own stories: the mother tending to her toddler's stubbed toe, the father coaxing his son down the slide, the baby inspecting a single grain of sand on the tip of her finger. When, and where, did their stories begin?

TWO MONTHS

The first time I realized I was pregnant, I was standing in the aisle of a supermarket.

A week before, I'd taken one of those extra-early pregnancy tests ("Our Most Accurate Results . . . Five Days Sooner!" read the box), and it had come up negative. Now I was leaning over the sushi case, debating my dinner options—Spicy tuna roll? Salmon and cucumber?—when John, standing beside me, said almost off-handedly, "Hey, did you ever get your period?" I was so persuaded by the authority of the slim pink wand and its austere single line that I had forgotten all about that more traditional indicator. John and I turned slowly to look at each other, a cartoon of dawn-ing awareness. "Oh my god," I said. Suddenly, a beer-and-sushi run to Whole Foods acquired a weighty new significance. We started our shopping trip all over again, pushing our cart back to the entrance of the store. Now we would be buying, and eating, for three instead of two, and we were determined to do it right.

Almost immediately, things got complicated. Halfway down the condiments aisle, I paused in front of the jars of peanut butter. Peanuts and other legumes are a healthy source of protein, I knew—but hadn't I read that eating nuts during pregnancy could predispose your child to allergies? We moved on. Rolling to a stop in front of the expansive dairy case, we hesitated again. Everyone knows that cheese is a good source of calcium, important for building strong bones. But I'd heard that soft cheeses, or those made with raw milk, may carry the bacterium listeria, which could cause birth defects or miscarriage. Onward to the chilly realm of the seafood department, where glassy-eyed fish regarded us from their bed of ice. I had a vague recollection that the fatty acids in fish promoted fetal brain development—but also that some fish contained pollutants like mercury and PCBs.

So it went, as we made our halting way through the store. In a flash, the world of food had been divided into good and bad, harmless and hazardous, and it was up to us to make the correct choices. Our dilemma was framed in the terms of modern science, yet it felt curiously primeval: we had been returned to the state of hunter-gatherers, figuring out for ourselves what was safe to eat. Instead of inspecting a mushroom on the forest floor, we were standing in the fluorescent aisle of a supermarket, scrutinizing a package's fine print. Baffled, John and I looked at each other over an empty cart.

No activity of everyday life is so instantly changed by pregnancy as eating. New worries about food safety turn ordinary meals into minefields. Waiters and counter clerks become subject to interrogation: Is the cheese pasteurized? Is the fish cooked through? Is the egg still runny, the meat still pink? More daunting still is the notion that what you eat becomes the very stuff of which your child is made. At times this idea is expressed with almost comic literalness—as when, during my first pregnancy, I signed up to get weekly emails from a parenting website. Each Saturday a message would appear in my inbox informing me of my fetus's current size, invariably described in terms of something edible. At nine weeks, I learned, my fetus was as big as a grape; at seventeen weeks, a turnip; at nineteen weeks, a "large heirloom tomato." The implication was clear: your baby is what you eat, and your baby had better not be the shape of a Twinkie.

Over the course of those nine months, I never quite got over the stunned anxiety I felt first in the bright, cold aisles of the supermarket. I did my best to eat healthily, but I continually second-guessed my choices, berated myself for slipups, worried that I was missing some crucial nutrient. All that angst and confusion come surging back when I learn that I'm pregnant a second time. Once again, eating is no longer a simple bodily function, much less a pleasure to be savored; it's a series of fraught choices, an act with grave consequences, committed three times a day. The questions tug at me every time I open a kitchen cabinet or gaze into the refrigerator: Is it true that what pregnant women put in their mouths can have an effect on their fetuses? What do we know about what's good and what's bad? And who can help me make sense of it all?

On a warm August morning in the second month of my second pregnancy, John and I head downtown for our initial doctor's appointment. In the examining room we're greeted by my obstetrician, a brisk, friendly woman who delivered my first baby. I hop up on the table and she places a probe on my stomach. On the monitor beside me, an elongated shape wavers into view; it looks, I can't help thinking, like a kidney bean. "Congratulations!" she says. "You're about seven weeks pregnant." She snaps off the ultrasound machine and leads us into her office, where she hands me a folder full of photocopied handouts. Most of them are about food. I barely have time to look at them before she begins asking questions.

"How many servings of fish do you eat a week?" she asks. "How many servings of dairy? How many servings of whole grains?" I stumble over my answers (how much is a "serving," anyway?), and feel like I'm failing a test. "You should gain three to five pounds in your first trimester, and about a pound a week after that, for a total of twenty-five to thirty-five pounds," she continues. She pauses, and looks me in the eye. "Now, let's go over food safety." I get out my pen and start taking notes: listeria, mercury, PCBs, toxoplasmosis . . . The sound of her voice fades away as the roar of a sudden fear breaks over me like a wave.

"A ham sandwich," I blurt. The OB looks up inquiringly.

"I'm sorry?"

"I ate a ham sandwich at a picnic two weeks ago." Oh god, I think, what have I done? Cold cuts can carry listeria! Those sandwiches were out in the sun for hours! The roar grows louder; I feel my face growing hot and my palms getting sweaty.

"If you haven't experienced any symptoms, you and your baby are probably okay," the OB says. She smiles at me, not unkindly. "But you have to be very careful."

Doctors (and a multitude of interested others) have long paid close attention to what pregnant women put in their mouths. Along with this attention have inevitably come attempts to influence the nature of the child by manipulating the woman's diet. "Because people knew so little about the fetus and how it was formed, they concentrated on what they could see and control," says Barbara Luke, a professor of obstetrics and epidemiology at Michigan State University who has studied the history of maternal diet during pregnancy. "If you look at the dictates given to

14

pregnant women over the centuries, a lot of them have to do with what they eat."

This was true of the ancients: "Pregnant women should above all else avoid repletion," or overeating, declared the second-century physician Galen, who noted that "servant girls and other poor women" are not "stuffed beyond due measure with food," yet they "reach their term easily, go into labor easily, and bring into the world a large and well-nourished baby." He concluded, a bit righteously: "Let that be a lesson to pregnant women." Other authors of ancient medical texts chimed in with warnings against foods that are heavy, sweet, acidic, or "flatulent," and urged pregnant women to consume a "moderate" and "wholesome" diet.

Even the God of the Old Testament got into the act: in the Book of Judges, the angel of the Lord appears before the woman who is to become Samson's mother. "Drink no wine or strong drink, and eat nothing unclean," warns the angel, "for lo you shall conceive and bear a son." For their part, generations of midwives and other traditional healers offered their pregnant patients counsel about eating that was one part common sense, one part superstition: a woman should eat more as her pregnancy progresses; a woman should balance foods whose essence was thought to be "hot" with those considered "cold."

The era of scientific advice to pregnant women about their diets began in the late nineteenth century, when researchers turned their attention to the subject. Ludwig Prochownick, an obstetrician in Hamburg, Germany, was the first to publish a study on diet during pregnancy, in 1889. Like many doctors of the time, Prochownick was concerned about the dangers of delivering big babies. Nutritional deficiencies in many women's own childhoods had left them with contracted pelvises; in an age before safe caesarean sections, a large fetus in a small pelvis could lead to grievous injury or death for both mother and child. Prochownick's solution was to place pregnant women on a strict diet: high in protein, low in calories and especially in carbohydrates, with very little salt and very little to drink. Based on his observations of three women, he claimed that his program produced smaller babies who were easier to deliver.

Prochownick's diet caught on among obstetricians in this country, and throughout the early decades of the twentieth century, they offered their own variations on what a pregnant woman should eat: cream of tartar mixed with lemonade, herring roe, cow's milk (two quarts every twenty-four hours), "the flesh of young animals," such as lamb and veal. And, of course, what they *shouldn't* eat: sweet foods, salty foods, and "the coarser vegetables," according to one Philadelphia physician. Like Prochownick, American doctors worried about difficult deliveries, so they instructed their patients to gain very little weight, often fifteen pounds or less. "It was believed that the baby would be fine, no matter what the pregnant woman ate," says Barbara Abrams, a professor of public health at the University of California, Berkeley, and an expert on maternal diet during pregnancy. "The reigning view was that the fetus was a 'perfect parasite,' taking from the mother all the nutrients it needed."

As the century went on, avoiding excessive weight gain became a preoccupation of many doctors, and so of their patients. "In the 1940s and 1950s, women were warned in very strong terms not to gain too much weight," Abrams tells me. "My mother used to starve herself before her visits to the obstetrician. In the middle of an Ohio winter, she would show up at the doctor's in sandals and a summer dress so that she would weigh less, because she was so afraid that the doctor would yell at her." Such strict enforcement of limits on weight gain started to loosen in the

1960s, when several large studies demonstrated that women who received adequate nutrition during pregnancy produced healthier babies. Doctors began to encourage their patients to "eat to appetite," using cues from their bodies, not numbers on a scale, to determine their diet.

This detente was short-lived. In recent years, as if the ghost of Ludwig Prochownick has returned to haunt us, doctors and scientists have again become focused on excessive weight during pregnancy. The numbers give them reason for worry: nearly two-thirds of American women of childbearing age are overweight; one in five women who gives birth in the United States is obese. A 2009 study found that up to 73 percent of U.S. women fail to follow guidelines for recommended weight gain during pregnancy, with most overweight women gaining too much.

As in Prochownick's day, there are concerns about difficult deliveries; overweight pregnant women are more likely to experience birth complications, and more likely to require a C-section. But today experts have other worries as well, borne of a new awareness that the intrauterine environment provided by a woman who is overweight may cause problems for the fetus well before birth. A number of recent studies have determined, for example, that the offspring of overweight or obese women are more likely to have birth defects (which, despite their name, have their origin at conception or early in gestation). A 2007 study published in the *Archives of Pediatric and Adolescent Medicine* found that the incidence of some defects was twice as high among the children of obese mothers. A 2009 study, the largest of its kind in the United States, reported that women who were overweight but not obese had a 15 percent increased risk of delivering a baby with certain heart defects.

Perhaps even more troubling is the notion that the prenatal

conditions provided by a woman who is overweight or obese when she becomes pregnant, or who gains an excessive amount of weight during pregnancy, may in turn promote the development of obesity in her child: what scientists call the "intergenerational transmission" of obesity. A 2007 study of 1,044 mother-child pairs, conducted by researchers at Harvard Medical School, found that greater weight gain by a woman during pregnancy was associated with a heavier child at age three. Women who gained more than the recommended amount of weight (that is, twenty-five to thirty-five pounds)—and even those whose weight gain was *within* the recommended range—had four times the risk of having an overweight toddler than women who gained less than the guidelines advise. A study by the same team, published the following year, suggested that this relationship persists into the offspring's adolescence: compared to the teenagers of women who had moderate weight gain during pregnancy, those of women who had excessive weight gain weighed more and were more likely to be obese.

Of course, children could well share eating habits, or a genetic predisposition to obesity, with their mothers—so how can we know that the prenatal environment is exerting an influence? Some researchers have found a clever way around this problem: they have compared children born to obese mothers with their siblings born after the mothers have had successful antiobesity surgery. Same mother, different intrauterine environment. In a 2006 study, published in the journal *Pediatrics*, researchers found that the children gestated by women postsurgery were 52 percent less likely to be obese than siblings born to the same mother when she was obese. Though these children had inherited their formerly obese mothers' genes, they were no more likely to be obese than the general population. A second study by the same group, published in 2009, found that children born after their mothers lost weight had lower birth weights and were three times less likely to become severely obese than their older siblings.

"The two groups of siblings are different, physiologically speaking, from one another," says John Kral, a professor of surgery and medicine at SUNY Downstate Medical Center in New York, who co-authored both papers. "The bodies of the children who were gestated after their mothers had weight-loss surgery process fats and carbohydrates in a healthier way than do the bodies of their brothers and sisters who were conceived when their mothers were still overweight." In effect, their metabolisms have been normalized by their prenatal experience. It may be that the intrauterine environment is as important as genes or family eating habits in passing on a tendency to obesity, says Kral. If that's so, he adds, "an obese woman who loses weight before getting pregnant is making an investment in her offspring's future health."

The mechanisms by which a pregnant woman's excessive weight could "program" her child's future size are not yet understood. Perhaps the intrauterine environment produces lasting changes in the fetus's body composition—its proportions of fat and lean body mass. Perhaps it persistently alters the functioning of the fetus's pancreas, which produces the sugar-processing hormone insulin. Indeed, John Kral's 2009 study found that the bodies of children born to women while they were still obese used insulin less effectively. Or perhaps it affects the way the fetus's appetite and metabolism function later on—possibly resetting the offspring's satiety point, so that more food is required to feel full. The 2009 Kral study also found that the offspring of still-obese women had less-optimal levels of leptin and ghrelin, two hormones that regulate appetite.

Animal studies suggest one further possibility: it may be that the food choices women make during pregnancy influence the later preferences of their offspring. In a 2007 experiment published in the *British Journal of Nutrition*, Stephanie Bayol and her colleagues at the Royal Veterinary College in London fed groups of pregnant and lactating rats either rodent chow alone or rodent chow along with generous quantities of junk food. "We actually went to the supermarket and picked out foods that humans eat: potato chips, jelly doughnuts, chocolate-chip muffins," Bayol tells me. "The rats especially liked the marshmallows we offered them."

After the rats gave birth, the researchers gave their ten-week-old pups a choice of rat chow or junk food. The offspring exposed to junk food in utero were 95 percent more likely to overeat than those whose mothers had eaten only rat chow, consuming an average of 22 percent more calories a day. They left their protein-rich chow almost untouched while they gorged on sweets, growing more than 25 percent fatter than their fellows. The pups' exposure to sugar, salt, and fat in the womb and in maternal milk, Bayol theorizes, affected the development of their brains' reward centers, priming their appetite for sugary, salty, fatty foods. Are there implications here for humans? Bayol thinks so. "We need to consider the possibility that women who consume junk food during pregnancy will make their children more likely to eat such foods themselves, and more likely to become obese," she says.

After hanging up with Bayol, I try to sort out the uneasy feelings that have settled in my stomach. Already, women are bullied into unhealthy eating habits by images of unattainable slenderness. Already, food has been moralized by the fear of gaining weight: carrot sticks and low-fat yogurt are "good," cookies and chocolate are "bad." Now, it seems, food poses a threat not only to one's waistline but to the health of one's future child. It doesn't take much imagination to see that this research could be used to add maternal guilt to the heavy emotional freight already borne by food. I think of an actual 2007 headline I spotted in *U.S. News & World Report*: "Don't Eat That, or Your Child May Grow Up Fat." Once again we're told that pregnant women are a danger to their fetuses, each bite they take a time bomb on a fork.

But there's another way to think about eating during pregnancy: as an act of sharing, even of teaching. Research suggests that more mature fetuses can experience tastes and smells in the womb; by seven months, the fetus's taste buds are fully developed, and its olfactory receptors appear to be functional. The flavors of the food a woman eats find their way into the amniotic fluid, which is continuously swallowed by the fetus. Babies seem to remember, and prefer, these familiar tastes once they are out in the world. In a 2001 experiment conducted by Julie Menella, a psychobiologist at the Monell Chemical Senses Center in Philadelphia, a group of pregnant women was asked to drink carrot juice during their third trimester; another group of pregnant women drank water instead. Six months later, the women's infants were offered cereal mixed with carrot juice, and their facial expressions were videotaped while they ate. The offspring of the carrot-juice-drinking women consumed more carrot-flavored cereal than babies who had not been exposed to the stuff before birth, and appeared to like its taste more.

Indeed, studies conducted with a variety of mammalian species demonstrate that young animals prefer flavors they encountered during gestation and lactation. Young rabbits devour the aromatic juniper berries their mothers ate during pregnancy; wild mice pups feast on fennel like that consumed by their dams; baby lab rats love the chocolate, rum, and walnut flavors fed to their mothers by experimenters. Such preferences are highly adaptive, Menella tells me. "Mothers are giving information to their offspring through what they consume during pregnancy and breastfeeding, telling them, 'This is what's good and safe for us to eat,'" she says. Among humans, such messages may be even more profound, bearing meaning not only about safety, but about culture. The characteristic flavors and spices of particular cuisines are likely introduced before birth, a prenatal initiation into one of culture's most powerful expressions: food. "Women are often advised to eat bland foods during pregnancy, but the way our food tastes is an important part of the world our children will be entering," says Mennella. "Through the foods a pregnant or lactating woman eats, she's educating her offspring in the flavor principles of her culture."

Analyses of amniotic fluid and breast milk have detected the presence of flavors as varied as garlic, curry, cumin, fenugreek, mint, and vanilla. In the Alsace region of eastern France, licorice-flavored anise is traditionally used to flavor candies, cookies, and drinks. In a 2000 experiment, Benoist Schaal, director of the European Center for Taste Science in Dijon, assessed the reactions of newborn babies to the smell of anise. The infants whose mothers consumed anise-flavored food and drink during pregnancy showed a preference for anise on the day they were born, and again later, on their fourth day of life. Babies whose mothers did not eat anise during pregnancy reacted neutrally or with outright aversion. "When a baby is born, he is not a blank slate," says Menella. "He has already been shaped by a rich array of sensory experiences that we are only now beginning to understand."

I ponder Menella's words one August evening when it's too hot to cook, and John and I have once again ordered in from our favorite Thai place around the corner. As I dig into my Penang curry with chicken and green beans, I'm thinking not about calorie counts or nutrient quotas, but about the distinctive diet to which my future child is already being exposed: the exuberantly polyglot menu of a twenty-first century Manhattan dweller with a world of tastes—Indian, Mexican, Chinese, Italian-American pizzeria—just down the block.

Given medicine's long-running preoccupation with women who eat too much during pregnancy, I'm surprised to learn that much of what we know about the long-term effects of maternal diet on offspring comes from a group of pregnant women who had little—far too little—to eat. In the autumn of 1944, the darkest days of World War II, German troops blockaded western Holland, turning away all shipments of food. The opening of the Nazis' siege was followed by one of the harshest winters in decades, so cold that the water in the canals froze solid. Food became scarce, with many Dutch surviving on just five hundred calories a day, a quarter of what they consumed before the war. As weeks of deprivation stretched into months, some resorted to eating tulip bulbs. By the beginning of May, the nation's carefully rationed food reserve was completely exhausted. The specter of mass starvation loomed—and then, on May 5, 1945, the siege came to a sudden end, when Holland was liberated by the Allies.

The Hunger Winter, as it came to be known, killed some ten thousand people and weakened thousands more. But there was another population that was affected: the forty thousand fetuses in utero during the siege. Some of the effects of malnutrition during pregnancy were immediately apparent in higher rates of stillbirths, birth defects, low birth weights, and infant mortality. Others would not be discovered for decades. Dutch epidemiologist Tessa Roseboom and her colleagues at the Academic Medical Center in Amsterdam have examined more than eight hundred individuals who were conceived and born during the siege, people now in their sixties. I call her to hear more about the fate of the children of the Hunger Winter.

"Our studies show that people whose mothers were pregnant during the siege have more obesity, more diabetes, and more heart disease in later life than individuals who were gestated under normal conditions," she tells me. "Their exposure to undernutrition in the womb appears to have had long-lasting effects on their health." These individuals' prenatal experience of starvation seems to have changed their bodies in myriad ways: they have higher blood pressure, poorer cholesterol profiles, and reduced glucose tolerance, a precursor of diabetes. Roseboom has found that the timing of the nutritional deprivation during pregnancy matters: the risk of diabetes is especially high among people exposed to malnutrition in mid-to-late gestation, while the risk for heart disease is three times higher in people whose mothers were starved very early in pregnancy.

"The deprivation experienced by pregnant women during the Hunger Winter was extreme," Roseboom concedes. "But in a sense the event presents an ideal chance to test the proposition that maternal diet during pregnancy can have effects lasting well into offspring's lives." The beginning and end of the siege were clearly demarcated: from October 1944 to May 1945. Everyone in the region blockaded by the Germans was affected: young and old, rich and poor. And the conscientious Dutch kept detailed medical records on each one of the country's citizens. From what one researcher has called "a tragic experiment of opportunity," scientists have gleaned a clue that nutrition in the womb may exert effects lasting for many years after birth.

As startling as they are, Roseboom's results come as no surprise to David Barker. Almost thirty years ago, Barker, a British physician, noticed something odd on a map. The poorest regions of England and Wales were the ones with the highest rates of heart disease. Why would this be, he wondered, when heart disease was supposed to be a condition of affluence—of sedentary lifestyles and too much rich food? Puzzled, Barker decided to investigate. He and his team searched for old birth records all over England, in lofts, sheds, garages, boiler rooms, and flooded basements. At last, in East Hertfordshire, they found a trove of thousands of turn-of-the-century records, jotted down by traveling nurses who visited each home where a baby had been born. Then Barker tracked down more than fifteen thousand of those individuals, now elderly, and compared their weight at birth (a crude measure of the quality of nourishment they received in the womb) with their health in later life.

The relationship was unmistakable: people who weighed less at birth had a higher risk of heart disease in middle age. "We were amazed by these early results," Barker has written. "The weights of babies and infants recorded long ago in cottages and terraced houses, measured by the light of candles and lanterns, using the simplest of weighing scales, were predicting heart disease fifty years later."

Barker's findings, first published in 1989, were not popular with his fellow physicians and researchers. At the time, scientists were making strides in documenting the relationship between heart disease and adult lifestyle factors like smoking, lack of exercise, and a fatty diet. There was also great excitement around the identification of genes associated with the development of heart disease. Barker's notion that the condition could be traced back to an individual's experience in the womb was met with profound skepticism, even ridicule. Still, Barker took hold of the idea and would not let it go, devoting the next decades of his career to plumbing the possibility he had uncovered. So closely was Barker identified with this notion—and so far did others want to stay from it—that it became known as the "Barker hypothesis."

Over time, however, Barker's dogged persistence began to change people's minds. In the years since his first study of the Hertfordshire birth records appeared, similar results have been reported by dozens of other scientists, studying a variety of different populations: in Europe, Asia, Australia, Africa, and North America. The mass of evidence in support of the idea, though not yet conclusive, is substantial enough that many original skeptics have become believers. Janet Rich-Edwards is an epidemiologist at Harvard Medical School who set out to disprove Barker's theory, using information from the Nurses' Health Study, her own long-running examination of more than 120,000 American nurses. "I was initially very skeptical of the idea of fetal origins," says Rich-Edwards. "I had the same bias as everyone else in the field of public health, which is that your current risk factors determine your odds of developing disease—not something that happened when you were a fetus." But, she says, "there's nothing like your own data to change your mind." Rich-Edwards was sure that once she took account of the nurses' adult lifestyles and socioeconomic status, the relationship between low birth weight and cardiovascular disease risk would disappear. "But the association barely budged," she says. "And this same study has been repeated at least twenty-five times now. It's one of the most solidly replicated findings in the field." The Barker hypothesis has gradually gained wide acceptance, and Barker is now far from its only champion.

I'm curious about this man who defied conventional wisdom so stubbornly for so long. I meet Barker one morning in the lobby of a hotel in midtown Manhattan, where he has come to attend a meeting of perinatal experts. Barker, seventy-one, is short and stocky, with an intensity that can veer into dry humor or curmudgeonly impatience. We sit down in one of the hotel's empty conference rooms, pitchers of ice water on the round table before us.

Why, I ask him, would undernutrition in the womb result in heart disease later?

"One explanation is that fetuses are making the best of a bad job," Barker replies. "When nutrients are scarce, they divert nutrients towards the really critical organ—the brain—and away from other organs, like the heart and liver." This keeps the fetus alive in the short term, he says, but the bill comes due later in life, when the heart, deprived early on, becomes more vulnerable to disease.

"But I believe that's not all that's going on," Barker continues. "I think that fetuses are actually taking cues from the intrauterine environment and tailoring their physiology accordingly. They're preparing themselves for the kind of world they will encounter on the other side of the womb." The fetus adjusts its metabolism and other physiological processes in anticipation of the environment that awaits it, Barker says, whether it's the parched desert of Ethiopia or the bounteous aisles of a Manhattan Whole Foods.

And the basis of the fetus's prediction is what its mother eats. In Barker's telling, the meals a pregnant woman consumes constitute a kind of story, a fairy tale of abundance or a grim chronicle of deprivation. This story imparts information that the fetus uses to organize the body and its systems, an adaptation to prevailing circumstances that facilitates its future survival. Faced with severely limited resources, Barker notes, a smaller-sized child with reduced energy requirements will have a better chance of living to adulthood. The real trouble comes when pregnant women are, in a sense, unreliable narrators: when fetuses are led to expect a world of scarcity and are born instead into a world of plenty. This is what happened to the children of the Dutch Hunger Winter, Barker says, and their higher rates of obesity, diabetes, and heart disease are the result. Bodies that were built to hang on to every calorie found themselves swimming in the superfluous calories of the postwar Western diet.

On a massive scale, that's what is happening now in developing countries like India, as people whose mothers survived on meager rations grow up in a world where Western-style food has become widely available and where McDonald's opens a new outlet every eight hours. Research by Barker and others has found that babies who are born small, then rapidly gain weight, are at particularly high risk of heart disease later. "In these countries we are seeing an epidemic of coronary heart disease, which will soon become the most common cause of death in the world," Barker tells me. "Today's underweight babies are going to be tomorrow's victims of heart disease."

Although Barker wasn't aware of epigenetics—the science of how genes are turned on or off, up or down—at the time he began his investigations, we now know that epigenetic mechanisms help explain how a pregnant woman's diet can have lasting effects on her offspring. One of the ways a gene's expression can be altered is through the action of a chemical cluster known as a methyl group. By attaching itself to a gene, a methyl group can turn that gene on or off. Especially while offspring are in utero, what their mothers do—or eat—can "methylate" particular genes in the offspring, changing how those genes behave. This was demonstrated in 2003 in a striking experiment performed by Duke University scientists Randy Jirtle and Robert Waterland.

The researchers started out with two groups of agouti mice, a strain of mice with yellow fur, fat bodies, and a predisposition to diabetes and cancer. A few decades ago, we would have assumed that these mice, and their descendants, were stuck with such characteristics: it's in their genes, after all. But now we know that agouti mice have the appearance and physiology they do because a certain gene, the agouti gene, is being continuously expressed. In these mice, the agouti gene has been activated, and so, Jirtle thought, perhaps it could be deactivated.

He fed one group of pregnant mice a regular diet. The other group of pregnant mice received a diet rich in methyl donors—nutrients capable of making methyl groups available in the body. The pups born of the first group of mice looked just like their parents—fat and yellow—and had a tendency to develop the same health problems. But the pups born of the mice who ate an enriched diet during pregnancy were different. They had brown fur, they were slender, and they were more likely to be healthy, without their parents' predispositions to disease. Their DNA hadn't changed—they still carried the agouti gene—but their epigenomes had, and it was prenatal nutrition that had flipped the switch. In ways that scientists are just beginning to understand, maternal diet during human pregnancy can also modify the fetus's epigenome in ways that affect its future health and well-being.

As I leave the cool dimness of Barker's hotel for the bright, hot, honking street, I'm thinking about my first baby's birth weight. I remember the card, stuck in his hospital bassinet, which bore his vital statistics: Theodore Paul Witt, born at 8:55 P.M. on 12/4/05, length 21 inches, weight 8 pounds 6 ounces. Like the note attached to a foundling left on a doorstep, or pinned to the coat of Paddington Bear, these few numbers were the sum of what we knew about him. Could his weight at birth really tell me anything about his future? Barker would say yes, that the relationship between birth weight and heart disease is found up and down the scale, not just at the extremes. "An eight-pound baby will have a lower risk of heart disease than a seven-pound baby, who will have a lower risk than a six-pound baby, and so on," he says. (Barker notes, however, that babies who are much heavier than average at birth also have an elevated risk of heart disease later in life. "This is one race where you want to come in right smack in the middle," he says.)

There are many potential causes of low birth weight: the baby may have been born early or have been one of a set of multiples; the mother may have smoked during pregnancy, or her placenta may not have functioned properly. But one major contributor to low birth weight is inadequate nutrition, and Barker's research has found poor nutrition to be widespread among women of childbearing age, even those who can afford to eat well. The Southampton Women's Survey is his ongoing study of more than twelve thousand young women living in the south of England. Its examination of women's health habits before, during, and after pregnancy has found that "only a small proportion of women planning a pregnancy follow the recommendations for nutrition and lifestyle," and that some of these unhealthy habits persist into pregnancy: nearly half of the women surveyed who became pregnant, Barker found, reported eating fewer than five fruits and vegetables per week.

We still have much to learn about the long-term effects of mothers' diets on their offspring. Still, one thing is sure: the notion of the fetus as a perfect parasite, entirely unaffected by its mother's diet, has joined the growing castoff pile of historical myths about eating during pregnancy. We now know that the fetus is indeed influenced by what the pregnant woman consumes—though how, and how much, remains the subject of fierce debate.

*　*　*

In my own mind, that debate has come to center on a single image, a vision that arrives to tempt me each day around eleven in the morning (lunchtime comes early for pregnant women). I dream about tuna salad on rye. In my avid imagining, the rosy tuna is bound by creamy mayonnaise and invigorated by the crunch of celery; it bears a fluted lettuce leaf and a fat slice of late-summer tomato. Actually eating such a sandwich would seem to be a simple act. Fish is low in fat and high in protein and iron; humans have been eating fish with evident enjoyment and without apparent harm for many thousands of years. In my carefree prepregnancy days, I ate a sandwich like this at least a couple of times a week. But, of course, eating fish is no longer a simple act, especially for pregnant women. In my thwarted fantasy of a tuna-salad sandwich are all the factors—the cultural pressures, the political calculations, the often-confusing scientific findings—that have made eating during pregnancy such a fraught activity.

Pregnant women's trepidations about eating fish stem, of course, from the fact that much of the world's fish supply is contaminated with methylmercury, a chemical produced by industrial processes and by natural events such as volcanic explosions and forest fires. In adults, mercury is rapidly absorbed into the bloodstream from the stomach and the intestine; it then readily enters the brain, where it accumulates and begins wreaking its neurotoxic effects. An early demonstration of the harm mercury can inflict came in 1956, when dozens of people living in the Japanese fishing village of Minamata began suffering strange symptoms. The victims became numb in their hands and feet, losing the ability to grasp small objects or to walk without stumbling; they developed difficulties hearing and seeing, and some were gripped by severe convulsions or fell into comas. At long last, the source of what had come to be known as "Minamata disease" was identified: a chemical factory was dumping its wastewater into Minamata Bay. The mercury in the factory's waste was absorbed by the bay's fish, which made up a large part of the villagers' diet.

The people most severely affected by the Minamata disaster, however, were those individuals whose mothers ate the contaminated fish while they were in utero. Some of these victims, now grown, traveled to Stockholm in 1972. There, at the first United Nations Conference on the Human Environment, they showed the world the wages of developmental mercury poisoning: slurred speech, deafness, blindness, mental retardation, and cerebral palsy.

Recognition that mercury could harm the fetus even at lower levels of exposure was slow in coming. In 1986, researchers in New Zealand reported that children subjected to moderate-to-high doses of mercury via their mothers' consumption of fish during pregnancy scored lower on developmental and cognitive tests at ages four and six. In 1997, scientists working in the Faroe Islands off the coast of Denmark found that children prenatally exposed to mercury from their mothers' diets exhibited deficits in memory, attention, and language at seven years of age. And in 2000, the National Academy of Sciences released a report, commissioned by Congress, assessing the dangers of mercury exposure. "The population at highest risk is the children of women who consumed large amounts of fish and seafood during pregnancy," the report concluded, adding that "the risk to that population is likely to be sufficient to result in an increase in the number of children who have to struggle to keep up in school and who might require remedial classes or special education."

At last, in 2001, the Food and Drug Administration issued an official advisory to pregnant women, warning them to avoid eating such fish as shark, swordfish, and tilefish, and to reduce their overall consumption of seafood. Women listened: a study by researchers at Harvard Medical School found that in the months following the FDA advisory, pregnant women reduced their fish consumption by about 18 percent, or 1.4 servings of fish per month (a serving was defined as three to five ounces of fresh fish and three to four ounces of canned tuna).

The confounding push-pull of eating during pregnancy was well illustrated by what happened next: a new round of research emphasizing the importance of nutrients found in seafood, called omega-3 fatty acids, to fetal brain development, and the consequent hazards of eating too *little* fish during pregnancy. "Omega-3 fatty acids are crucial for the development of the central nervous system, and low seafood intake during pregnancy could lead to a deficiency of these nutrients in the fetus," Joseph Hibbeln tells me. Hibbeln, a scientist at the National Institutes of Health, led a 2007 study published in the leading journal the *Lancet* that found that low levels of seafood consumption during pregnancy were associated with increased risk of low verbal IQ, social and communication problems, and poor fine-motor skills in children six months to eight years of age. (These "suboptimum outcomes" were found among the offspring of women who ate fewer than twelve ounces of seafood per week during pregnancy; twelve ounces is less than half of what the pregnant women in the Harvard Medical School study were eating following the FDA advisory, suggesting that many pregnant women are consuming less than ideal quantities of fish.) Adding to the bewildering blizzard of messages were industry groups promoting the safety of seafood and activist groups raising alarms about the continuing danger of mercury contamination—and the news media, covering it all in sensational detail.

What's a pregnant woman to do? I dive deep into the research, its strong currents pushing me first one way—fish is good!—and then the other—mercury is bad!—and finally surface with a seemingly obvious conclusion: eat fish, lots of it, just not the mercury-laden kind. This is harder than it sounds. Fish, to me, has meant a pink slab of tuna steak or a creamy slice of swordfish (or that coveted tuna-salad sandwich). Now I set about getting acquainted with the ocean's other inhabitants: small fish at the bottom of the food chain, shellfish that trawl the ocean floor, vegetarian fish that dine on plants instead of fellow sea creatures. Soon, I'm finding all kinds of ways to eat fish: Sardines on buttered toast for breakfast (delicious with scrambled eggs). A snack of herring on crackers with a dab of Dijon mustard. For dinner, shrimp sautéed in butter with garlic and cayenne pepper. Catfish, breaded and broiled golden. Tilapia on the grill, with some mango salsa to perk up its admittedly anemic flavor. And anchovies slipped into everything: puttanesca sauce, pesto, spaghetti with garlic and oil. As I tuck into a snowy filet of flounder, squeezed with lemon and speckled with pepper, I think to myself: eating carefully can actually taste really good.

Once I remember this, I'm surprised that I ever forgot. So unforgiving are the attitudes on eating during pregnancy that they can twist even a previously uncomplicated relationship to food. All my life, food had been for me an unmitigated pleasure: by turns familiar comfort and exciting adventure, physical and emotional sustenance served up on a plate. Through some quirk of metabolism, I was spared the weight watching that tormented many of my female friends. As kids, my sister and I were so string-bean skinny that, at the old-fashioned summer camp we attended in Maine, the camp director enrolled us in the Milk and Cookies Club. After lunch, while the other campers were napping in their bunks, Sally and I and a few similarly scrawny preteens were led back to the dining hall and fed a second helping of dessert. The cook's specialty was something called a mud pie, which was a vanilla cupcake scooped out with a spoon and filled with chocolate sauce. As a treatment strategy for underweight children, it was indefensible—but, licking the chocolate from our lips as we crept back into the hushed cabins, we weren't complaining.

It wasn't until I became pregnant that I experienced food the way so many others do, as a malign temptation, full of tricks and snares. The more I read and thought about eating during pregnancy, the more its perversities seemed an exaggerated version of a more general societal derangement over food. In fact, the book I find most useful during my second pregnancy isn't about pregnancy at all. It's the *The Omnivore's Dilemma*, by food writer and social critic Michael Pollan. As Pollan describes it, Americans have grown increasingly disconnected from food production, food preparation, and even from the experience of eating itself, becoming reliant on processed foods manufactured in industrial facilities and consumed in the car or in front of the television. Obsessed with health and slenderness, but hooked on salt, sugar, and fat (not to mention speed and convenience), we have developed what Pollan calls a "national eating disorder," a state of confusion and anxiety about what to eat. The "omnivore's dilemma," in his telling, is the perplexity created by the fact that humans can eat just about any food we wish. When the question "What's for dinner?" has an almost limitless number of answers, and when, as in the United States, we lack strong food traditions to guide us, we fall back on fad diets and sketchy science.

American women can't help but bring these distorted ideas about food into pregnancy. In a nation where cookies are the size of dinner plates and a bowl of spaghetti can feed a family of four, we don't know what appropriate portion sizes are anymore. In a country where the prompts of appetite are overruled in bouts of binging and dieting, we can no longer use feelings of hunger and fullness as reliable guides. And in a culture in which thin is the only acceptable size, it's hard for us to take pleasure in our changing shape. Indeed, the new pressures of pregnancy often exacerbate already-unhealthy attitudes toward eating: our belief that making a meal of processed food is much quicker and easier than cooking, for example, or our assumption that nutrient-supplemented "edible foodlike products," in Pollan's phrase, are an adequate substitute for real food.

So unsure is our relationship to food, Pollan points out, that we rely on the edicts of government and the media to tell us what to eat. Over the past three decades, such commands have become especially numerous for pregnant women. In 1980, the U.S. Food and Drug Administration, for the first time, advised pregnant women to reduce their intake of caffeine. Warnings about bacteria-bearing cheese, parasite-infected meat, and toxin-laden fish soon followed. Newspaper and magazine articles, and pregnancy guidebooks in particular, drove home the dire message: for pregnant women, eating lunch is an operation as delicate and dangerous as defusing a bomb. Many women can quote from memory a notoriously punitive passage from *What to Expect When You're Expecting*, first published in 1984: "Before you close your mouth on a forkful of food, consider, 'Is this the best bite I can give my baby?'" it reads. "If it will benefit your baby, chew away. If it'll only benefit your sweet tooth or appease your appetite, put your fork down."

In the face of such scolding, the bountiful buffet of the omnivore withers to encompass just a few harmless (and often flavorless) foods; the problem becomes not one of too many choices, but too few. The American fear of food becomes a full-blown phobia, and pregnant women develop what might be called a nine-month eating disorder, afraid to ingest anything that isn't doctor or expert approved. For too many of us, eating during pregnancy represents not a departure from unhealthy habits, but the elevation of these habits to a new extreme.

It doesn't have to be this way, says nutritionist Karin Michels. "Pregnancy is actually an ideal occasion for shaking up the way you think and feel about food," she tells me. "You're already forced to pay more attention to what you eat during pregnancy, so you can take advantage of that heightened awareness to make positive changes that will last after the nine months are over." I've traveled to Boston to meet with Michels, an associate professor at Harvard Medical School and an expert on eating during pregnancy. Her office is lined with heavy tomes about diet and nutrition, but my eye keeps being caught by something else: a row of tiny stuffed animals atop her computer monitor. Michels herself is intently serious, especially when she talks about her laboratory research on the role of prenatal factors on the development of cancer in later life. She grows more animated, however, when we start talking food. "Healthy, wholesome food can be so tasty," she exclaims. "You don't have to deprive yourself of great food when you're pregnant. It's all about what you buy and how you prepare it."

Michels feels so strongly about this that she wrote a book intended for pregnant women, *The Gift of Health*, which I read on the train ride from New York. It was the first book on diet during pregnancy, I realized, that didn't make me feel like I'd been caught with my hand in the cookie jar. Michels presents eating during pregnancy as joyous affirmation, not anxious rejection or grim self-denial. In place of defensive eating, she advocates using pleasure, of all things, as a guide. (Admittedly, not the pleasure of a package of Oreos or a Big Mac with fries, but the pleasure, still genuine, of nutty whole-grain bread or sweet young vegetables.) "I'm doing it for my baby" is often uttered with a sigh of resignation, but it could just as easily be a slogan of liberation: permission to try a new cookbook, check out a neighborhood bakery, sample the bounty at the farmers' market. Why not? I'm doing it for my baby!

In coming years, this positive posture toward eating during pregnancy may be reinforced by research produced by scientists like David Williams. Williams, a principal investigator at the Linus Pauling Institute at Oregon State University, is testing the notion that certain foods consumed during pregnancy can help prevent offspring from developing diseases like cancer later on, providing them with chemoprotection from illness that may last for their whole lives. "We have evidence that a number of chemicals that are protective against cancer and other diseases are capable of crossing the placenta and are available to the fetus," Williams tells me. "So, my question was: Can we provide these chemicals through the maternal diet, and in this way provide disease protection for the offspring? And at least in the animal model we've developed, that's exactly what happened."

Williams's experiments have shown that the offspring of mice who ingest a phytochemical derived from cruciferous vegetables like broccoli, cabbage, and brussels sprouts during pregnancy were much less likely to get cancer, even when exposed to a known carcinogen. The same is true of the offspring of mice who were given green tea during pregnancy. After they were weaned, the offspring in William's experiments never encountered these protective chemicals again, yet their exposure during pregnancy and lactation was enough to shield them from cancer well into maturity. I ask Williams if he can imagine a day when pregnant women are prescribed a dietary supplement that will protect their future children from cancer. "It's not science fiction," he replies. "I think that's where we're headed."

Right now, sitting in Karin Michels's office, I need a more immediate kind of assistance. Feeling sheepish, I go ahead and ask her anyway: "Could you help me with my grocery list?"

Back in New York, the sun is baking the sidewalk and softening the asphalt of the street; inside the supermarket, the chilled air feels good on my bare arms. I commandeer a cart and start walking briskly toward the produce section, Karin Michels's list clutched in my hand.

"Make your plate colorful," she had advised, and the late-summer bounty makes that easy: into the cart go red tomatoes (full of beneficial carotenoids), orange yams (rich in vitamin A), and green spinach (lots of calcium and iron). Eating a variety of fruits and vegetables of different colors, Michels had explained, ensures that you're getting a full complement of micronutrients. Now I add a couple of avocados, full of B vitamins, plus potassium; a few bunches of broccoli, rich in riboflavin and folic acid; and handfuls of cherries, an excellent source of vitamins A and C. Passing over my usual salad greens—the relatively low-nutrient romaine—I scoop up some arugula, which is high in vitamin K and is one of the few foods aside from fish that is rich in omega-3 fatty acids.

Next stop is the bread and cereal aisle. I look for products with whole grains, and check the ingredients list, as Michels had instructed: the shorter, the better. I find a toasty-brown loaf made of just whole-wheat flour, water, yeast, and salt, and plunk it into my cart; boxes of whole-grain crackers and cereal follow. Now I'm on a roll. Off the shelf comes a big bottle of olive oil, a healthy, monounsaturated fat. A few pounds of beans, which provide protein without the saturated fat of red meat. A dozen eggs rich in choline, a B-complex vitamin that's a building block for the neurotransmitters necessary for learning and cognition. And some nuts for snacking: almonds, hazelnuts, and walnuts, all rich in nutrients like vitamin B-6, and in minerals like copper, magnesium, and manganese. Nuts appear to be safe to eat during pregnancy, Michels told me; if I had a family history of severe food allergies, she said, I might avoid highly allergenic peanuts just to be safe. I even brave the seafood department, buying a half-dozen creamy scallops, low in mercury and high in vitamin E.

At last, list exhausted, I push my cart into the checkout line, where a candy display temps the idling shopper. I've read with interest the emerging research about the effects of eating chocolate during pregnancy. A study conducted by scientists at Yale University and published in 2008 reported that, compared to pregnant women who ate less than one serving of chocolate a week, those who consumed five or more servings of chocolate each week during their third trimester had a 40 percent lower risk of developing the dangerous high blood pressure condition known as preeclampsia. Another study, by researchers at the University of Helsinki in Finland, found that mothers who ate chocolate every day during pregnancy reported that their infants showed less fear and smiled and laughed more often at six months of age.

I'd like to have a happy baby, and I'd rather avoid preeclampsia. I have another motivation in mind, however, as I reach over and drop a bar of extra-dark chocolate into my cart: it just tastes good.

THREE MONTHS

On a crisp evening on the edge of autumn, I linger in the doorway of a downtown restaurant, exchanging goodbyes with my dinner companions. I don't want to go, I realize, don't want to leave the cozy company of my friends or the warm light pooling on the sidewalk. At last I wrest myself away, and as I head into the chilly dark, pulling my jacket tightly around me, I feel it again: a new and acute sense of vulnerability. Hurrying down Seventh Avenue toward the subway, crowds streaming past, I realize that I'm warily scanning the faces of the strangers who walk by, and that I've put a protective arm across my belly, at three months just beginning to swell. Being pregnant has made me vigilant, on guard for threat; I feel alert and instinctive as an animal, ears perked and nose twitching.

Earlier in the day, I had talked by phone with Dan Fessler, an associate professor of anthropology at UCLA. "It's not your imagination," Fessler told me. "Biologically speaking, pregnant women actually *are* more vulnerable." So that the pregnant woman's body doesn't reject the fetus—which is, after all, half unrelated to her—her immune system is suppressed, making her and her fetus more susceptible to pathogens. To counter this vulnerability, Fessler theorizes, humans have evolved a set of behaviors that emerges during pregnancy, a kind of emotional and psychological immune system that compensates for a diminished physical one.

Experiments performed by Fessler and his colleagues show that pregnant women are more likely than nonpregnant women to react with disgust to a variety of sources of infection or contamination, such as bodily excretions—especially early in pregnancy, when the risks posed by pathogens are greatest. More provocatively, his research suggests that while pregnant, women become more xenophobic: more distrustful of strangers and more favorably disposed toward members of their own group. In a study of 206 pregnant women published in 2007 in the journal *Evolution and Human Behavior*, Fessler and two coauthors reported that pregnant women in their first trimester reacted more negatively than women further along in pregnancy to an essay critical of the United States written by a foreigner, and more positively to an essay praising the United States written by an American. Similar research by psychologists at the University of St. Andrews in Scotland has found that pregnant women have a stronger preference than do nonpregnant women for vigorous, healthy looking faces over sickly, unhealthy looking ones.

The motivation behind such evolved protective mechanisms remains below our awareness, even as these feelings steer us away from potential harm—say, from outsiders who might be carrying diseases against which we lack immunity. "The reason we experience these reactions today is that the response protected our ancestors," Fessler said. "These emotions allowed our forbears to survive long enough to produce offspring, who in turn passed the same sensitivities on to us. We often respond to today's world with yesterday's adaptations."

To hear Dan Fessler tell it, the vigilance I feel on this city sidewalk is old and ingrained, bred into my DNA over eons of evolution. But surely, I tell myself, with all the safeguards of civilization—our penicillin and our pasteurized milk, our seatbelts and bike helmets and bottles of sunscreen and Purell—we're now well beyond the body's blunt defenses. We're buffered from danger, insulated from threat. That's when I look up and see them: two columns of light, soaring into the cloud-streaked sky. The powerful spotlights are rising from the site of Ground Zero, a couple of miles away at the tip of Manhattan. How could I have forgotten? Today is the anniversary of the terrorist attack of September 11, 2001. On this night each year, ghostly reminders of the fallen Twin Towers appear. I stop at the edge of the subway steps and stare.

When I first moved to New York a decade ago, the towers helped me find my way around a dauntingly huge and complex city. Emerging from the subway, I would look for their profiles on the horizon, then steer myself by their unchanging compass. Back then I lived in the West Village, where the towers were visible from almost every vantage point, an imposing but oddly comforting presence. My navigational assistance, and some of my ease of being in the city, vanished one September morning in the middle of breakfast. John buzzed up from the street, where he'd gone to walk the dog. "Come on out here," he said, sounding puzzled. "You need to see this."

Feeling at once annoyed and curious, I trotted down the stairs and out the front door. A knot of people was gathered in the middle of Greenwich Street, gaping and pointing at the sky like extras in a B movie. Following the direction of their gaze, I saw a dark, unfathomable hole in one of the towers; even as I struggled to take in this sight, I watched a plane strike the second tower. From a distance of thirty blocks, it looked like a shower of sparks.

John and I climbed the stairs in stunned silence. Back in the apartment, it suddenly seemed very important to exchange my flip-flops for a pair of sneakers: I wanted to be ready to run. The urge seemed irrational until I turned on the television and saw hundreds of high-heeled shoes abandoned on the streets around the fallen towers, tokens of refinement and sophistication that had suddenly become an impediment to survival. The attack on the World Trade Center was a brutal reminder that modern life still bristles with threats, that danger can arrive on a sparkling late-summer morning while you sip your tea. Now that I'm pregnant, this unsettling fact strikes me with new force. Is it true, as I feel it to be, that pregnant women and their fetuses are especially vulnerable?

At 8:46 AM on that bright, blue-sky day in September, there were tens of thousands of people in the vicinity of the World Trade Center: commuters spilling off trains, waitresses setting tables for the morning rush, brokers already working the phones on Wall Street. About 1,700 of these people were pregnant women. When the planes struck and the towers collapsed, many of these women experienced the same horrors inflicted on other survivors of the disaster: the overwhelming chaos and confusion, the rolling clouds of potentially toxic dust and debris, the heart-pounding fear for their lives.

As the catastrophe began to unfold, psychiatrist Rachel Yehuda was arriving for work at the Bronx Veterans Affairs Medical Center, about fifteen miles north of the Twin Towers. "I was leading a meeting at the center when I got a call from my mother, who lives in Florida," Yehuda tells me. "She had seen news of the attack on TV and wanted to know if I was all right." Yehuda and her colleagues located a television of their own and watched, aghast, as the awful events of that day took shape. "Of course, I was thinking about what the long-term reactions of the survivors would be," she says. Yehuda, who is director of the Traumatic Stress Studies Division at the VA and a professor at the Mount Sinai School of Medicine in Manhattan, is a leading expert on post-traumatic stress disorder, a condition that forces survivors of a traumatic event into a state of hypervigilance, assailing them with nightmares and panic attacks. In the course of her career as a PTSD researcher, she has worked mostly with Holocaust victims and Vietnam War veterans—people whose trauma happened far away and many years, even many decades, ago. As Yehuda watched in real time as tragedy struck her own city, she was already thinking about how to investigate its impact.

In the years since 9/11, Yehuda has coauthored more than a dozen articles about its effects on survivors, including several deeply intriguing studies of women who were exposed during pregnancy. "I wanted to look at this population in particular because I have a long-running interest in the transgenerational transmission of PTSD risk, or the handing down of a susceptibility to PTSD from parent to child," she explains. Yehuda encountered a vivid example of this phenomenon in 1993, when she opened the first clinic in the world devoted to the psychological treatment of Holocaust victims. She expected a flood of inquiries from people who had experienced Nazi persecution firsthand. She soon got a surprise: for each call her clinic received from a Holocaust survivor, it was getting five calls from their grown children. "Many of these members of the second generation had symptoms of PTSD,"

Yehuda says. They reported the same nightmares, the same panics, the same hair-trigger vigilance that their parents had. Yehuda's research confirmed that the offspring of parents with PTSD were more likely to develop PTSD themselves, even though they were no more likely to encounter traumatic events than other people.

How could this be? The traditional, psychoanalytically inflected explanation was that the younger generation was sensitized to trauma by growing up around survivor parents—hearing their stories, observing their struggles, enduring their silences. In the words of one commentator, the children of Holocaust victims bore "the scar without the wound." Yehuda, too, initially subscribed to this notion. "I was really convinced that the early experiences of the child, being bombarded for years by the psychiatric symptoms of the parents, accounted for the transgenerational effect we observed," she tells me. In the years after her clinic opened, she began to wonder if some other influence was at work—perhaps even before birth.

Yehuda's previous research had found that low baseline, or basal, levels of cortisol were a marker of vulnerability to PTSD. People with low basal levels of cortisol are more likely to develop PTSD following a traumatic event. Cortisol is a hormone released when the body is under stress; one of its functions is to stop the stress reaction once it has run its course. In people with low levels of cortisol, the body's heightened state of alarm, so necessary in the midst of a crisis, does not subside once the crisis is over. Yehuda's research had also established the striking fact that the offspring of people with PTSD have low basal cortisol as well. Is this commonality simply genetic? Or could it be passed down in utero? Nine-eleven provided an opportunity to find out.

Along with her coauthors, Yehuda enrolled thirty-eight women who were pregnant when they were exposed to the World

Trade Center attack in her study, measuring their basal cortisol levels and those of their infants when they reached one year of age. Those women who developed PTSD following 9/11 had low basal cortisol levels, and so did their babies, an effect that was most pronounced in infants whose mothers experienced the catastrophe in their third trimester. Since the children were only a year old, it didn't seem possible that the psychoanalytic explanation—that they were traumatized by their parents' stories—was the case here. "The particularly strong effects seen after exposure in the third trimester point to prenatal factors, rather than genetic or parenting factors, in the transmission of PTSD risk," says Yehuda. It looked like the mothers with post-traumatic stress disorder had passed on a vulnerability to the condition to their children while they were in utero.

A subsequent study of Holocaust survivors and their grown children provided another piece of supporting evidence. In this experiment, Yehuda found that the offspring were more likely to develop PTSD if their mothers, but not their fathers, had PTSD—suggesting, she writes, "that classic genetic mechanisms are not the sole model of transmission." In tracing the roots of PTSD, Yehuda tells me, "we need to be looking where we hadn't even considered looking before"—in the womb.

There's one more twist to this story. Post-traumatic stress disorder appears to be a reaction to stress gone very wrong, causing its victims tremendous unnecessary suffering. Jonathan Seckl, Yehuda's collaborator and a professor of molecular medicine at the University of Edinburgh in Scotland, suggests another way of thinking about PTSD. "What looks like pathology to us might actually be a useful adaptation in some circumstances," Seckl tells me. "In a particularly dangerous environment, the characteristic manifestations of PTSD—a hyperawareness of one's surroundings, a quick-trigger response to danger—could save someone's life." Other scientists who study prenatal stress have advanced the same conjecture: "We can speculate that extra vigilance and rapid shifts in attention could be adaptive in an environment full of danger or predators," writes Nicole Talge of the University of Minnesota. "In our own civilization, with no predators, and in which great premium is put in education on focus and concentration, extra vigilance and rapid shifts in attention can be maladaptive, and result in unnecessary anxiety and problems with attention." I'm aware that this theory is still speculative, but I find it oddly poignant. It would mean that even before birth, mothers are warning their children that it's a wild world out there, telling them: *Be careful.* It would mean that all of us, no matter how privileged or civilized, are never far from what Rachel Yehuda calls "the biology of survival."

At 4:31 on the morning of January 17, 1994, an earthquake hit Northridge, California, twenty miles northwest of Los Angeles. The quake, which measured 6.8 on the Richter scale, killed fifty-seven people and injured nearly twelve thousand others. At home in Laguna Beach, sixty miles away from the epicenter, was Curt Sandman, a professor of psychiatry at the University of California, Irvine. "Our house is suspended over a hillside, and the whole structure shook severely during the earthquake," Sandman tells me. "The sound it made was remarkable, like a train coming right through the house." After the quake's reverberations had stilled, Sandman found himself wondering how this event would affect pregnant women.

Stress during pregnancy is Sandman's area of expertise; indeed, at the time of the Northridge earthquake he was already tracking a group of pregnant women to learn more about how stressful life events affected birth outcomes. From this pool of subjects, he identified forty women who had experienced the earthquake at close hand. "The value of this set of circumstances was that we had assessed the women's psychological functioning before the earthquake, and that the earthquake itself was an event experienced by all these women at exactly the same moment," Sandman says. "Because we knew the gestational interval of each woman in the study, we could precisely examine the influence exerted by this shared event." Women who were in their first trimester when they experienced the earthquake, Sandman found, delivered their babies two weeks early on average—twice as early as women who were in their third trimester when the earthquake hit. Women who did not experience the earthquake during pregnancy, a group included in the study as a control, had the longest gestations of all. "There seems to be a critical period during pregnancy during which the woman is most susceptible to stress," Sandman tells me. "The earlier in pregnancy the earthquake occurred, the earlier the delivery."

His study adds to a body of research indicating that extremely stressful events can affect pregnancy outcomes. Like Sandman, a number of scientists have found that severe stress is associated with a higher risk of early delivery, or of having a baby with low birth weight. Pierre Buekens, dean of the Tulane University School of Public Health and Tropical Medicine in Louisiana, examined the birth outcomes of 301 pregnant women living in New Orleans and Baton Rouge when these cities were struck by Hurricane Katrina in 2005. Buekens and his coauthors found that women who'd had three or more "severe hurricane experiences," such as walking through floodwaters, sustaining significant home damage, or going without electricity for more than a week, were at a "markedly increased risk" of delivering low birth weight and preterm infants. A far larger study, carried out in Denmark, examined the records of more than one million women who gave birth over a twenty-four-year period. Women who had experienced the death or serious illness of a relative during or just before pregnancy were 16 percent more likely to deliver prematurely; when it was one of their own children who had died or become ill, the incidence of premature delivery increased by 23 percent.

Just how extreme stress exerts its effects on pregnancy is not yet clear, but scientists are investigating several possible mechanisms. Severe or chronic stress may constrict the pregnant woman's blood vessels, reducing the amount of oxygen and nutrients delivered to the fetus, or it may affect the functioning of the placenta in some other fashion. Increased levels of the stress hormone cortisol may interfere with fetal development; in a study by Miguel Diego of the University of Miami School of Medicine, pregnant women with high levels of cortisol measured in their urine and saliva were more likely to have fetuses who weighed less and were smaller when measured in midpregnancy. Or the woman's experience of great stress may trigger a cascade of other chemical changes in her body. Some scientists, including Curt Sandman, believe that such stress may activate a placental "clock" that determines the length of gestation, raising the risk that birth will be initiated early.

To be clear: these are not everyday hassles we're talking about, the wear and tear of ordinary life. Extreme stress is produced by events that threaten one's life or the lives of loved ones: the sudden death of a close family member, the diagnosis of a terminal disease, a natural disaster, a terrorist attack—or the experience of war. Indeed, one of the earliest scientific studies of the effect of extreme stress on the fetus had its roots in a conflict known as the Winter War of 1939. In November of that year, three months after the start of World War II, the Soviet Union invaded the small nation of Finland. The Soviet forces outnumbered the Finns four to one, but the Finnish troops' fierce determination and skillful fighting in freezing conditions held off the invaders for three and a half months. By the time it finally surrendered, Finland had lost twenty-five thousand soldiers, many of them men with young wives and small children.

Nearly four decades later, Matti Huttunen, a professor of psychiatry at the University of Helsinki, decided to look into the fate of children whose fathers had been killed. With a colleague, Pekka Niskanen, he identified 167 children whose fathers died while they were still in the womb, and compared them to 168 children whose fathers died during their first year of life. (Some of the fathers in their study died from causes other than the war.) He found significantly more cases of schizophrenia and behavior disorders among individuals who had lost their fathers prenatally—perhaps, he conjectured, because they had been affected during gestation by their mothers' grief. Huttunen published his results in 1978, urging more research into how events experienced by pregnant women, such as bereavement, might affect their offspring in utero. "The pregnant mother represents the critical environment of the fetus," he wrote, "just as the quality of early human relations represent the critical environment of the newborn."

One researcher who has followed Huttunen's lead is Dolores Malaspina, a professor of psychiatry and environmental medicine at New York University. Malaspina examined the health records of almost eighty-nine thousand people born in Jerusalem between 1964 and 1976. She found that the offspring of women who were in their second month of pregnancy during June 1967, the time of the Arab-Israeli Six Day War, were significantly more likely to develop schizophrenia as young adults. The effect varied by sex: females whose second month in the womb coincided with the conflict were 4.3 times more likely to develop schizophrenia than females gestated at other times, while males were 1.2 times more likely. Malaspina theorizes that these individuals' neurological development in utero was disrupted by stress hormones generated by their mothers' bodies in response to the war, and that the second month of pregnancy is a period of particular vulnerability.

After reading the literature of extreme stress for a while, I start to feel a little shellshocked myself. It's a relentless, almost Biblical litany of plagues: battles and floods and earthquakes and hurricanes. There's a reason these dramatic events dominate the research: the effects of stress during pregnancy are vexingly hard to study. Women who are under stress *during* pregnancy are often under stress *after* pregnancy as well, potentially affecting their parenting and making the specific effects of prenatal stress hard to identify. Women who are genetically predisposed to create stressful, chaotic lives may pass on this same tendency to their children, calling into question whether prenatal stress is the cause of the offspring's troubles or merely a side effect of mother and child's shared heredity.

Extreme events cut through such complications to create what scientists call a natural experiment. They affect everyone in a given community, the troubled and the psychologically healthy alike. They happen within an identifiable period of time, which can be matched up with pregnant women's gestational stage. Then they end, creating a discrete experience rather than an ongoing state of existence. Such circumstances make unexpected events the next best thing to a traditional scientific experiment, in which subjects can be randomly assigned to one condition or another. In other words, exactly what makes these experiments of nature so useful to fetal researchers is what makes them so frightening to the rest of us—that they can strike anyone, at any time.

On January 6, 1998, a freezing rain began to fall over a wide swath of Canada, the start of what was to become the worst natural disaster in the country's history. As temperatures plunged and the rain turned to ice, millions of people lost heat and electricity. Soon grocery stores ran out of bread; gas stations emptied their tanks; hardware stores sold out of generators and oil lamps. Banks and pharmacies served their customers by flashlight or candlelight, and hospital emergency rooms filled with people with bones broken from slipping on the ice.

One of the people caught in the ice storm was Suzanne King, an associate professor of psychiatry at McGill University in Montreal. "The ice froze on everything—on the streets, the sidewalks, even on the high tension lines, which toppled huge transmission towers like dominoes," King tells me. When she tried to clear the rain from her car windows, her windshield wiper snapped in two. King's husband was out of town on a business trip, so she was alone in caring for her two children, aged six and four. When the electricity in her house went out, she moved the family to her mother-in-law's, then to her younger child's preschool, as each place in turn lost power.

"We think we're so civilized and then we plummet right down to the bottom, to matters of survival," King says. "I was driving around town looking to buy water to drink, to buy fuel to cook with. It was all about the basics—food, water, shelter, staying warm. The hardest thing was not knowing how long it was going to go on." Finally, the rain stopped and the cold relented—but that wasn't the end of it for King. "A few days after things had returned to normal, I went to donate blood, since the hospitals were running low," she recalls. "The nurse took my blood pressure, and it was sky high. That's weird, I thought—and then I realized that my body was still stressed out from our experience during the ice storm. I immediately wondered: What is this doing to pregnant women and their fetuses?"

King's turn of thought was prompted by her psychiatric research, which focuses on risk factors for schizophrenia. She knew that maternal stress during pregnancy has been implicated in the development of this mental illness, and she saw in the ice storm an opportunity to observe the effects of such adversity in real time, following prenatally stressed offspring from birth onward. Within months King had initiated Project Ice Storm, sending out a survey that posed questions like: How many days were you without electricity? How often were you required to change residence during the ice storm? Did your family stay together for the duration of the ice storm? Was anyone close to you injured?

King and her colleagues began following more than 150 women who had been pregnant during the storm, administering questionnaires to the mothers and tests to their children at periodic intervals. Their earliest results showed that the more stressful events pregnant women encountered during the disaster, the lower were their babies' birth weights. Evaluations performed when the offspring were two years old revealed an association between prenatal stress and cognitive and language skills: the more severe the stressful events experienced during pregnancy, the poorer were their the toddlers' abilities. A third round of tests performed when the children were five and a half showed continued cognitive and language delays, as well as increased rates of attention and behavior problems, among children whose mothers endured high levels of hardship during the ice storm.

Now ten years old, the children of women who encountered great adversity during the ice storm have displayed differences at every stage from kids whose mothers had an easier time. To be sure, these differences are subtle—"variations in normal," as King puts it. Most of the high-stress children are doing well in school, and most score above average on IQ tests, albeit ten to twenty points lower than the low-stress control children. But the differences are present and persistent, a result that King finds surprising and dismaying. "The impact of ice storm stress on the offspring is greater than I thought it would be," she tells me. "I was surprised to find effects on the kids when they were two years old. And I thought for sure by the time they were five, or eight, the prenatal effects would have washed out and the postnatal environment would have taken over. But that's not what we've found." After following these children for more than a decade, King says, "I have a lot more respect for the impressionability of the fetus. Historically, people knew that it was a good idea to take special care of pregnant women. But in modern times, we've forgotten that." Especially in crisis situations, she says, "we need to see pregnant women and their fetuses as high priority."

Indeed, new evidence of the fetus's vulnerability to severe stress, along with the dramatic recent disasters of 9/11 and Hurricane Katrina, have focused emergency planners on the need to care for victims who are pregnant. (Katrina alone displaced at least ten thousand pregnant women.) Measures now being considered, or already being implemented, include evacuating pregnant women early when a hurricane or other natural disaster is predicted, providing information specifically for pregnant women in emergency briefings and public service announcements, offering telephone hotlines or face-to-face counseling to pregnant women who are victims of a disaster, and training emergency professionals to be aware of pregnant women's particular needs. Even the simple act of asking each woman accessing emergency services if she might be pregnant can make a positive difference, since her pregnancy may not yet be visible, and since the early months of pregnancy are crucial ones for the formation of the embryo. Such information allows women who are pregnant to be steered to appropriate medical and psychological resources, and can be used to create a registry that tracks pregnant women and their outcomes.

Pregnant women themselves can take steps to prepare for a crisis, like having copies of medical records and any necessary medication at the ready. Some researchers have advocated incorporating disaster planning into childbirth education classes. But even after reading a stack of journal articles on the subject, the notion of creating an emergency plan or putting together a disaster kit remains purely aspirational for me, a task I'm sure I'll take on someday, like organizing my closets or alphabetizing my books. Then comes the morning in late September when, late for a pediatrician's appointment, I'm racing Teddy in a stroller down Broadway. Suddenly, a poster on the side of a bus shelter brings me up short. "YOUR DISASTER *CLIENT MEETING. LUNCH. EMAILS.* PLAN IS MORE IMPORTANT. MAKE A PLAN," admonishes the sign, sponsored by the Department of Homeland Security. My own excuses would go something more like *PEDIATRICIAN. PRESCHOOL. PLAYDATE*, but I get the message, which feels placed there just for me. My household, I know, is woefully unprepared for an emergency as minor as a blown fuse. I'm at the opposite end of the continuum from those survivalists who could live for years on their stores of bulgur wheat and dried lentils: I live over a supermarket, which I visit at least once a day. My family could subsist on what's in our cabinets for twenty-four hours, tops, after which we'd be down to a jar of capers and a few animal crackers.

I did once try to put together a disaster kit; the attempt, it must be said, was not a success. This was back in the strange, spooked days after 9/11, when Homeland Security Secretary Tom Ridge advised us all to buy duct tape and plastic sheeting and when there was cachet in wrangling a prescription for the anti-anthrax drug Cipro. Getting into the spirit, I lugged home a case of water and filled a duffel bag with food: stoned wheat crackers, tins of anchovies, cans of white beans. "This is supposed to be a disaster kit, not a yuppie picnic," protested John. Over time, the items in the bag disappeared one by one: first the household's sole can opener was reclaimed for the kitchen, followed by the crackers, purloined to serve some unexpected guests. After a while, all that was left were the jugs of water, one of which sprung a hole and leaked, very slowly, all over my closet floor.

I vow that this time will be different. One night, after Teddy is in bed, John and I draw up our emergency plan, following the recommendations we find on the website of the American Red Cross. We choose two places to meet in case of an emergency (outside our apartment building and outside his office). We pick an "out-of-area emergency contact person" (my mom, who lives in Philadelphia). We decide on responsibilities (he'll get Teddy from preschool, I'll handle the dog—and the baby, once it comes). The next day, I set out to gather provisions for a disaster kit, again guided by the Red Cross: flashlights, a battery-powered radio, a first-aid kit.

My shopping feels, I find, like a funny kind of nesting; instead of collecting crocheted blankets and tiny onesies, I'm buying extra batteries and cans of soup (not to mention a designated can opener). The whole project reminds me of the hospital bag pregnant women are supposed to have ready to go, packed with slippers and bathrobe and a pillow from home. Both measures seem, in different ways, so inadequate to the immensity of the events they're intended to prepare us for. And yet, once my disaster duffel is packed full and zippered closed, I feel a gratifying sense of security, like a squirrel all stored up for winter.

Emergency preparations completed, I find myself wondering about more everyday kinds of stress during pregnancy. Earlier generations had strong views on the subject: the Victorians, for example, believed that women should refrain from virtually all activity during pregnancy, including housework, travel, and even social calls. In one popular domestic guide, *From Kitchen to Garret: Hints for Young Householders*, published in 1889, the chapter on pregnancy is titled "In Retirement." Pregnancy, author Mrs. Jane Panton counseled, is a time "when the mistress has perforce to contemplate an enforced retirement from public life." Today, attitudes have undergone an almost complete reversal. The notion that women should restrict their activities during pregnancy is regarded as old-fashioned at best, and patronizing at worst. How well a woman handles pregnancy is gauged not by how gracefully she submits to a new regimen of rest, but by how determinedly she adheres to her busy prepregnancy schedule.

During my first pregnancy, my days were even more frenetic than usual, as I hurried to get assignments completed and arrangements in place before what felt like the biggest deadline of my life. Running to catch the subway just as the doors closed, dashing across a crowded midtown street, pacing with nervousness before an important meeting with an editor, I did sometimes wonder if I should be taking it easier. This time around, I've got another slew of deadlines, plus a highly effective stress producer known as a three-year-old.

My hectic lifestyle puts me in good company these days. More than two-thirds of American pregnant women work, most of them full time, according to the most recent figures from the U.S. Census Bureau. Of those who are employed, 80 percent are still at their jobs in the last month before the birth; in the early 1960s, that number was only 35 percent. No doubt many women make this choice out of simple financial necessity: they need the income, and if they have time off they'd rather take it after the baby is born.

But there's something else going on here, too: the rise of an invincible icon I think of as the Pregnant Superwoman. This imaginary superhero never needs to sit down and take the weight off her feet, or catch her breath at the top of the stairs. She runs a marathon in her third trimester and works twelve-hour days until her water breaks. She's just like a nonpregnant mortal, in fact, only better, and she has become the unattainable standard against which women measure themselves during pregnancy.

My guess is that the Pregnant Superwoman got her start when women entered the workplace in large numbers in the 1960s and 1970s. Getting ahead required proving that nothing about being female—including the inconvenient tendency to get pregnant—would interfere with one's performance or one's commitment to the job. A woman could work just like a man, even when carrying thirty extra pounds of blood, fluid, tissue, and kicking, squirming fetus. The mythology of the Pregnant Superwoman got another unintended boost from the women's health movement of the 1970s. Feminist activists exhorted the medical establishment to treat pregnancy not as a disease but as a natural, normal condition—a worthy aim that was understood by some to mean that pregnant women required no special attention or treatment from the world at large.

Such attitudes linger on today, as does an expectation that women will not burden bosses and coworkers, or friends and family, with their needs during pregnancy. Over time external pressures like these became internal ones, as women began to take pride in not giving any ground to pregnancy, in not being weak or dependent. "I know that when I was pregnant, I liked being macho," Suzanne King, author of the ice storm study, confessed to me. "I'd stand on a chair, hammering a nail, and if my husband objected, I'd tell him, 'Don't treat me with kid gloves! I'm not an invalid!'"

Hearing yet another story of a woman who went into labor at her desk or in a business meeting, I wonder if we've taken all this too far. Our superhero stoicism during pregnancy might make life easier for spouses and bosses and coworkers. But how does it affect the pregnant woman herself, and her fetus?

It's clear enough that the extreme stress of a traumatic event—as well as chronic stressors like being poor and living in a crime-ridden neighborhood—can have negative consequences for a pregnant woman and her fetus. Whether that's true of more ordinary work stress is a scientific gray zone, filled with varying findings and opinions. One who believes that work stress may well be problematic is Calvin Hobel, an obstetrician at Cedars-Sinai Medical Center in Los Angeles and a pioneer in the study of stress during pregnancy. When he began his research on the impact of stress on preterm birth twenty-five years ago, he tells me, "A lot of people thought it was just ridiculous—stress wasn't even on the list of suspected causes of preterm birth. Now, it's at the top of the list."

He points to studies like one recently conducted by researchers at University College Dublin and published in the *British*

Journal of Obstetrics and Gynecology. Enrolling more than six hundred working women in Ireland, the study found that women who worked at physically demanding jobs had four times the risk of delivering an underweight baby than women whose jobs were less taxing. Women whose jobs were temporary had four times the risk of delivering a premature baby than women who hold permanent positions.

Hobel has worked to develop programs for disadvantaged pregnant women who suffer the stresses of poverty, racism, crime and domestic violence. But he also worries about his more affluent patients. "Some of the most stressed-out women I see are the lawyers and executives who continue to work long hours, all the way to the end of their pregnancies," he says. "Even when they're put on bed rest, they take their laptops and their phones into bed with them and keep working."

Ideally, says Hobel, a woman planning to have a child would work with her doctor to identify and reduce sources of stress before pregnancy. Once she's pregnant, her stress level would be evaluated each trimester. (Hobel's proposal has been seconded by the American College of Obstetricians and Gynecologists, which in 2006 recommended that all women receiving prenatal care be screened once a trimester for "psychosocial risk factors," including stress.) A comprehensive instrument for conducting such screenings has not yet been developed, Hobel tells me, but he offers to walk me through the elements such a screen should contain.

The "structured interview" he conducts with me begins with questions about basic safety and welfare: Have I been the victim of crime, domestic violence, or discrimination? Do I have the resources to meet my basic needs? Next come questions about big life events: Have I recently moved, changed jobs, or gone through a divorce? Have I experienced the serious illness or death of a family member, or a natural disaster or other trauma? These are followed by questions about relationships and social support: Do I have a loving relationship with a partner, and do I have close bonds with family and friends? Next he asks me about everyday stress: Do I feel burdened by work obligations, or by domestic responsibilities like caring for children or elderly parents? And finally, he probes my self-perception and perceived stress: Do I feel in control of my life? Do I feel that I'm a capable and worthwhile person? Reviewing my answers, Hobel and I agree that my stress level seems to be under control. Pregnant women whose stress level is high, he said, can be counseled by their doctors or referred to a mental-health professional.

Those with moderate levels of stress, like me, can take comfort from the work of another expert, Janet DiPietro. DiPietro, a developmental psychologist at Johns Hopkins University in Maryland, has a somewhat different take than Hobel: "Most pregnant women can stop worrying about stress hurting their fetuses," she says flatly. An enzyme in the placenta, she notes, breaks down the cortisol in a pregnant woman's blood, preventing most of it from reaching the fetus (though this mechanism can be swamped when extreme stress produces very large amounts of cortisol). And moderate stress, her own research has found, may actually be *good* for the fetus, accelerating its maturation.

Just as adults find that some amount of stress focuses their attention and pushes them to perform at their best, for fetuses, "mild stress may be a necessary condition for optimal development," DiPietro has written. In a study of ninety-four women and their offspring, published in 2006 in the journal *Child Development*, she and her coauthors found that women who reported modest anxiety and daily stress during pregnancy had children with better motor and mental development scores at age two.

62

This finding persisted even after controlling for other possible contributing factors, such as the anxiety and stress the women experienced postpregnancy. In another, more recent investigation, DiPietro reported direct biological evidence of this relationship between prenatal stress and accelerated development. The study of 112 mother-child pairs found that the two-week-old infants of women who experienced relatively more stress during pregnancy showed faster neural conduction—evidence of a more mature brain.

"Among its many functions, the stress hormone cortisol plays a role in the maturation of body organs, and exposure to relatively higher levels of this chemical could enhance the fetus's development," DiPietro explains. Women who experience periods of stress in their daily lives, she speculates, may effectively be giving their fetuses' nervous systems a beneficial workout, toning and conditioning fetal responses in a way that persists after birth. That's not to say pregnant women should run themselves ragged, DiPietro emphasizes: "I tell pregnant women: You should avoid stress because it's bad for *you*. Not because it's bad for the fetus, because I don't believe it is. Doing yoga, cutting back on your work hours—if it makes you feel better, that's a good thing. But don't worry that you're harming the baby."

Hobel and DiPietro's different takes on the effects of everyday stress during pregnancy mirror a split in the research literature: some studies suggest grounds for concern, while others conclude it's not a problem. To the question of whether such stress unfavorably affects pregnant women and their fetuses, we may have to settle for that profoundly unsatisfying answer: it depends. It depends on whether you work long hours, at physically demanding tasks. It depends on whether you're energized by a fast pace and a load of responsibilities, or exhausted by them. It depends on whether you have a sense of control at work, and a source of emotional support at home. DiPietro offers the useful concept of the "inverted U-shaped curve": most people, pregnant women and their fetuses likely included, do best when they have a moderate but manageable amount of stress in their lives. Aim for the fat middle of the inverted U, she says, rather than one end (too little stress) or the other (too much stress).

But such nuance is seldom found in the public discussion of stress during pregnancy, where fear and anxiety rule. When DiPietro talks or writes about the inverted U-shaped curve, she tells me, she knows to expect questions from pregnant women about how to achieve precisely the "right" level of stress—including queries from women who want to *add* the appropriate amount of stress to their apparently too-easy lives.

Media coverage of research on stress during pregnancy often only stokes such concerns. Take, for example, the international media frenzy that greeted a 2007 study by Dutch scientist Gouke Bonsel, a professor of social medicine and public health epidemiology at the Academic Medical Center in Amsterdam. "Women's Work Stress Harms Unborn Babies"! "Pregnant Women in High Stress Jobs Should Cut Working Hours"! "Working 32 Hours a Week or More in Pregnancy 'Is as Risky as Smoking'"! shouted the headlines.

What Bonsel actually found was that, in a study of 4,976 Dutch women, those who experienced high "job strain" while pregnant— that is, employment situations in which they had a lot to do, and not much control over how they did it—had a significantly higher chance of having an "excessively crying" infant who cried for three or more hours a day. In a subsequent study, published in the *American Journal of Public Health*, Bonsel and three colleagues examined 7,135 women during their first trimester of pregnancy. They found that a high level of job strain, and a work week of thirty-two hours or more, were each associated with somewhat decreased birthweight. Women who had both risk factors gave birth to babies who were five ounces lighter than average, a reduction similar to that found in women who smoke during pregnancy.

Reading Bonsel's scientific papers, I notice that his findings come with caveats generally overlooked in the hyperventilating media coverage: "We found that in general there is no reason to assume that working during pregnancy has a negative effect on birth weight," he wrote near the end of the second article. Employment during pregnancy, he noted, is mostly associated with *positive* outcomes; unemployed pregnant women are more likely to suffer from depression, and may give birth to babies with lower birth weights. And the unfavorable combination of the highest job strain and the longest workweek was found among only a small percentage of the women in his study. Still, Bonsel and his coauthors concluded, "our results suggest that reducing job strain and working hours in the initial stages of pregnancy may be beneficial among women with stressful full-time jobs."

Given that two-thirds of American women work during pregnancy, Bonsel's conclusion is a thought-provoking one—though a far cry from the assertion that "pregnant women should halve the amount of time they work or risk damaging their unborn child," as one newspaper described his findings. Such alarming warnings to pregnant women have a long history, I learn, in which science has played only a part.

On September 4, 1905, Curt Muller, the owner of Grand Laundry in Portland, Oregon, ordered laundress Emma Gotcher to work more than ten hours in a row. This violated an Oregon statute limiting women's work hours in laundries and factories; in the late nineteenth and early twentieth centuries, dozens of state legislatures passed such laws. Muller was cited and fined ten dollars, but he appealed—all the way to the United States Supreme Court, where he gave his name to one of the nation's most famous cases, *Muller v. Oregon*. Arguing his side before the Court was the brilliant lawyer Louis D. Brandeis, who assembled what came to be known as the Brandeis Brief: 113 pages of medical and sociological data documenting the harm caused to women by overwork, with an entire section devoted to its "Specific Evil Effects on Childbirth and Female Functions."

The brief presented a roll call of opinions from doctors, scientists, and government agencies. Declared Geo. Newman, M.D.: "An occupation requiring a woman to stand during the greater part of the day when continued up to within a few days or even hours of the time of parturition, must act to the detriment of the offspring, and there is less chance of the latter coming into the world fully grown, well-formed, and in good health." Asserted J. H. Bridges, M.D.: "The evils occurring in women" as a result of their work in textile factories "indirectly affect the more perfect growth of the child in utero, and dispose it when born more easily to become diseased." Reported the Massachusetts Bureau of Statistics of Labor: "The necessity for instrumental delivery has very much increased within a few years, owing to the females working in the mills while they are pregnant and in consequence of deformed pelvis." Noted Dr. F. B. Kane, of San Francisco: "Very many times my attention has been drawn professionally to the injury caused by the long hours of standing required of the saleswomen in this city, the one position most calculated to cause the manifold diseases peculiar to their sex, and direfully does Nature punish the disobedience of her laws."

Such testimony—the first time this kind of extralegal evidence had been accepted by the Court—proved persuasive to Justice David Josiah Brewer, author of the opinion finding against Curt Muller. "That woman's physical structure and the performance of maternal functions place her at a disadvantage in the struggle for subsistence is obvious," Brewer wrote. "This is especially true when the burdens of motherhood are upon her. Even when they are not, by abundant testimony of the medical fraternity continuance for a long time on her feet at work, repeating this from day to day, tends to injurious effects upon the body, and, as healthy mothers are essential to vigorous offspring, the physical well-being of women becomes an object of public interest and care in order to preserve the strength and vigor of the race." Therefore, Brewer concluded, "her physical structure and a proper discharge of her maternal functions—having in view not merely her own health, but the well-being of the race—justify legislation to protect her from the greed as well as the passion of man."

At the time it was handed down, the *Muller* decision was celebrated as a victory for female workers. A Portland newspaper praised this "epoch in American civilization" for "removing woman from the category of beasts of burden whose labor may be exploited by her industrial masters without restraint." Yet laws like Oregon's, with their emphasis on protecting female employees' baby-making capacities, came uncomfortably close to treating women like farm animals, valued for what their bodies could produce. Louis Brandeis made this all too clear a few years after *Muller*, in an argument in favor of a minimum wage for women working in factories and stores. "Does anybody doubt that the only way you can get work out of a horse is to feed the horse properly?" he asked. "Does anyone doubt that the only way you can get hens to lay is to feed the hens properly?" Working women must be treated well if they are to fulfill their all-important maternal function: "Experience has taught us that harsh language addressed to a cow impairs her usefulness," Brandeis pointed out. "Are women less sensitive than beasts in these respects?"

I find myself musing on the meaning of these comparisons of women to animals, which appear with striking frequency in my research—from Brandeis's early twentieth century argument to a 2006 newspaper article, which credits hormones for "the cow-like tranquillity of pregnant women." As someone who, before children, lived mostly in her head—the kind of person who'd rather read a book about something than actually do it—I have in fact been startled by the insistent physicality of pregnancy, by the urgent impulses and animal fears it stirs in me. It can seem like a useful corrective to our sleek modern lives to be reminded that we're not ambulatory computers, after all, but warm-blooded mammals who gestate fetuses and give birth and make milk. During pregnancy, the body and its considerable powers can't be denied. And yet I'm also keenly aware that the likening of pregnant women to animals has often been used to suggest that they are no longer thinking beings, but creatures of pure instinct—not fully human, and certainly not full citizens or workers.

Indeed, laws limiting the activities of female workers came over time to be seen as discriminating against women rather than protecting them, and in the 1960s and 1970s, American lawmakers tried to ensure that employers treated women, including pregnant women, just like any other employees. Title VII of the Civil Rights Act, passed in 1964, prohibited employment discrimination based on sex, and the Pregnancy Discrimination Act, passed in 1978, added specific prohibitions against discrimination on the basis of pregnancy, childbirth, or related medical conditions.

Limits on the work performed by pregnant women soon found another form: so-called "fetal protection policies," which claimed concern for the fetus rather than the pregnant woman. The legality of such policies was tested in 1991, when the case *Automobile Workers v. Johnson Controls* came before the U.S. Supreme Court. A group of employees and their unions had sued Johnson Controls after the company banned all women, except those with a doctor's note stating that they were infertile, from certain jobs making car batteries. Were any of these female workers pregnant, Johnson executives explained, exposure to the lead-containing batteries could harm their fetuses.

Justice Harry Blackmun wrote the opinion finding in favor of the workers. "Concern for a woman's existing or potential offspring historically has been the excuse for denying women equal employment opportunities," Blackmun wrote, pointing to *Muller v. Oregon*. "It is no more appropriate for the courts than it is for individual employers to decide whether a woman's reproductive role is more important to herself and her family than her economic role." That choice, he declared with the firmness of a gavel rap, is "hers to make."

Such rulings have settled, at least for now, the matter of pregnant women's legal right to work, while leaving the cultural debate about women's activities and ambitions during pregnancy as active as ever. It's a subject that has roiled American society at least since Curt Muller ordered Emma Gotcher to work overtime, and I find that my own feelings about it are divided, too. I know I don't want Supreme Court justices, or even psychologists and public health experts, to tell me how long or how hard I should be working while pregnant. I have no desire to return to the sti-

fling prohibitions of the Victorian age, or the paternalistic pro-
tections of the Progressive Era. And yet it seems to me that, in
our own time, pregnancy comes with a set of expectations that
is oppressive in its own way. Women today are given to understand that pregnancy shouldn't slow them down in the slightest;
at the same time they're barraged with warnings that too much
stress could harm their fetuses (a kind of modern Brandeis Brief).
That's a lot of cultural baggage to bear, at a time when women
have enough additional weight to carry. I find myself daydreaming about a pregnancy free of these burdens, concerned only with
the optimal experience of woman and fetus. What would such a
pregnancy look like?

It might look something like my day today. It's Saturday,
which means little when you're a freelance writer on deadline: the
morning would usually have found me with half a mind on Teddy
at play and half a mind on the laptop open on my kitchen table
(with predictably half-brained results). But today, after a solid
week of work, I decide to give myself some time off—or really,
some time *on*: fully engaged with my son, putting worries about
word counts and editors' comments aside for now.

This is often easier said than done, I know. Financial necessity, employers' inflexibility, and women's own commitment to
their work may prevent them from taking time away from their
jobs while pregnant. Some women, and their doctors and employers, have come up with creative ways around these obstacles. A
few companies, for example, have begun offering an arrangement
called "future leave": employees put aside a portion of their regular paycheck each month, which is then used to pay for a leave
during pregnancy or at another time of need. Other women continue working during pregnancy but do so from home, skipping
the stress of a commute. And some pregnant women build time

for rest into their schedules simply by rearranging their work hours. Calvin Hobel, the California obstetrician with many high-powered clients, used to implore his patients to take some time out during their hectic workweeks, to no avail—until he hit on a solution that some could accept: they work one day of the week-end, and take Wednesday off as a kind of midweek break.

Since I'm daydreaming, I can take the notion of time off even further: antenatal leave, or formalized leave offered to women during pregnancy. Not surprisingly, European countries have already institutionalized it. In France, women are required to take two weeks of leave in advance of their due date, and can take as many as six weeks. In Finland, the seventeen and a half weeks of maternity leave each woman receives can begin as early as eight weeks before her baby is expected to arrive. When women in this country are offered antenatal leave, a recent study shows, birth outcomes improve on at least one important measure. Sylvia Guendelman, a professor of public health at the University of California, Berkeley, compared a group of sixty-two women who took antenatal leave to 385 women who worked up until their babies' births. (California is one of only five states in the nation to offer paid pregnancy leave to its residents, through an insurance program similar to ones that provide unemployment benefits.) Those who took the time off, Guendelman found, were four times less likely to require a caesarean delivery. "Women who take time off from work seem to come to the delivery room more rested and relaxed," she says. "But in this country we're not very generous to pregnant women—we don't have a culture that encourages pregnant women to take care of themselves." If for no other reason, Guendelman adds, granting antenatal leave makes good economic sense, since C-sections cost more and require more recovery time for the mother.

After I help construct (and reconstruct, following a building-wide collapse) a block castle with Teddy, John takes over and I take off for a yoga class with a friend, also pregnant. Usually I find yoga too slow moving and ponderous for my taste, like a roomful of people imitating sloths. But my interest has been piqued by a small but intriguing literature on the effect of yoga and relaxation exercises on pregnancy outcomes. A study of fifty-eight pregnant women, published in 2005, found that participating in relaxation exercises decreased participants' feelings of stress, and significantly reduced the levels of the stress hormone cortisol found in their blood. Several studies of prenatal yoga, all carried out in India, reported that pregnant women who practiced yoga for an hour a day felt less stressed, were less likely to experience preterm labor or birth complications, and had babies with higher birth weights.

The yoga instruction I've signed up for is a special prenatal class, and as the participants drift into the room, taking off their shoes, spreading out their mats, I marvel at the variety in the shapes of their bodes: from the willowy woman, newly pregnant, whose abdomen is still almost concave, to the woman so heavily pregnant with twins she holds on to her belly like she's grappling with a medicine ball. As the class begins I find that I'm paying unaccustomed attention to my own body as well: to my breathing, already a bit more labored, and to the way my sense of balance has begun to alter, vertebrae shifting back to compensate for the new weight in front. The class ends with the group in a circle, each woman in turn sharing a thought or feeling about her baby to be. When the hour is up I feel loose-limbed and relaxed, and full of a feeling of kinship with my pregnant classmates. On the way out we all smile at each other, as if we're keepers of a shared secret.

Scientists call this social support, and they've found that it's associated with positive pregnancy outcomes. Women who have robust social support, especially from a number of different sources—friends, family, coworkers, neighbors—tend to have babies with higher birth weights. Some obstetricians are even trying to build such support into pregnant women's routine medical treatment by organizing group prenatal care, in which women can form friendships with each other even as they have their questions answered and their blood pressure checked. "Of course, many women already have social networks in place in their lives," obstetrician Hobel notes. "We ask women, 'Who are your friends? Can you lean on them a bit more while you're pregnant, can you ask them for help?' The research suggests this can really make a difference." I mention these findings to my friend, and together we enjoy a guilt-free postyoga chat over lunch.

When I get home John is strapping Teddy into a stroller, the only way we can get him to nap these days. I volunteer for the walk, and as I nudge the stroller's wheels onto the wide paved path in Riverside Park, I find myself thinking again about that indomitable icon, the Pregnant Superwoman. Her myth, I realize, is perpetuated in part by pregnant women themselves: she reassures us with her cheerful invulnerability, her bulletproof resistance to all the changes—physical and emotional and logistical—that come along with having a child. Her fearless independence allows us to evade the fact that pregnancy and childrearing *do* make us more dependent on others, in the heat of a disaster and in the heart of everyday life. In short, the Pregnant Superwoman embodies the insistence that pregnancy doesn't change anything, a fable that may hold as much appeal for pregnant women as it does for spouses and employers. She tells us that, for nine months more, we can hold off the tidal wave of change we know is coming—

but at the cost of slighting the needs that have already arisen, and ignoring the changes that are already here.

Back in the apartment, I peek under the blanket covering Teddy's stroller: he's out, eyes closed, snoring softly. Usually I'd rush to my computer, hoping to dash off another few paragraphs or return a bunch of overdue emails. But today, I decide, I'm taking off the Pregnant Superwoman cape. Spread out on the couch in my home office, it makes an inviting spot for an afternoon nap.

FOUR MONTHS

"I would love a glass of wine," I say.

And then sigh. "But I'll just have a some water, thanks." The hostess gives me a sympathetic smile as she pours; the bubbles in the sparkling water fizz and pop in a wan imitation of festivity. It's a chilly night in October, and John and I are at a dinner party at a friend's apartment. Looking at the faces gathered around the table, the flushed cheeks and shining eyes, I can tell the evening is hitting its sweet spot: that moment when jokes sound funnier and food tastes better, when the conversation swells and the light grows golden. At four months, there are still times when I forget that I'm pregnant, but there's no more conspicuous reminder of my condition than having to step outside the warm circle of social ritual—in this case, sharing a bottle of red with a bunch of friends. My chair is drawn up to the table like the others, but I feel distinctly set apart.

Maybe it's the children's table I'm seated at. It seems perverse that on the cusp of an experience that epitomizes adulthood—becoming a parent—pregnant women are expected to renounce so many grown-up pleasures. The steaming hot bath, the blue-veined cheese, the glistening slice of raw fish: they get their zing from a touch of danger, of decay, and in the case of alcohol, of poison. All these pose some potential risk to the fetus, and so on our way to maturity pregnant women are returned to the nursery, to sliced-cheese sandwiches and tepid baths and sparkling apple juice with dinner. On top of that, pregnancy has multiplied the minor afflictions that would usually have me reaching for a drink at the end of the day: my back aches, my feet hurt, my mind is full of worries and preoccupations. A spicy glass of Pinot sounds pretty good right now.

But not for me. The moment I learned I was pregnant, I became a born-again teetotaler, a fanatical practitioner of clean living—swearing off not just alcohol but caffeine, cold medicines, headache remedies, anything that comes in a pill or capsule. At times, feeling some physical discomfort—and pregnancy brings more than its share—I'll open up the medicine cabinet and start to reach for one of the bottles crowded on the shelves, before realizing they're all off limits. This abstemious regimen is largely self-imposed—my OB gave me a list of drugs that are safe to take during pregnancy—but most of the pregnant women I know act the same way. It's an odd turnabout. For many of us, pregnancy is the first time in our adult lives that we're not "on" something—birth-control pills, sleeping aids, antidepressants, the Advil or the antacids we usually down like candy. Living amid a chemical cornucopia, we've become premodern, pharmaceutical Luddites shunning so much as a cup of herbal tea.

I'm sure it wasn't always this way. My own mother, pregnant with me, drank her way through a trip to the British Isles, and still maintains that the pub-sign slogan "Guinness Is Good For You" applies to expectant women as well. Other pregnant women she knew took tranquilizers, sleeping pills, even amphetamine-laced diet pills to keep their weight down. I find myself disturbed by these stories, but in truth I don't know exactly where my feelings about the use of alcohol and medications during pregnancy come from—only that they are viscerally, unreasoningly powerful. Sipping my chaste glass of sparkling water, I wonder: How did we get in just a few decades from the easy attitudes of my mother's generation to the anxious ones of my own—and whose view is right?

I resolve to find out. Pregnancy and its enforced remove from the familiar can be an opportunity, I know, to see things afresh, from a more clear-sighted (not to say sober) perspective. It's a small comfort I cling to as the noise of the party reaches a giddy crescendo around me.

Early the next morning I head into my home office, mind clear and thoughts already percolating; some time later John shuffles, moaning, into the kitchen to pour himself a cup of strong coffee (another adult privilege I've regretfully relinquished). I've got a stack of histories of pregnancy and alcohol on my desk, and what I read leaves me dizzy with surprise. Pregnant women were once permitted, even encouraged, to drink alcohol by their doctors. In the nineteenth century, physicians prescribed champagne as a treatment for morning sickness, and brandy with soda as an appetite stimulant. Well into the twentieth century, alcohol was viewed as an all-purpose remedy that soothed pregnant women's nerves and fortified them for the rigors of labor. It was believed that alcohol would relax the uterus following amniocentesis, and even arrest labor that had begun prematurely. (This theory was championed by a Cornell University scientist who used it on his own wife when she began having contractions at seven months.)

Wide-eyed, I read accounts of obstetricians giving their patients intravenous drips of pure alcohol; of pregnant women so drunk that nurses had to strap them into their beds; of maternity wards that smelled, recalled one doctor, "like a bar."

"At the better hospitals, women in labor might be handed a vodka and orange juice," Janet Golden tells me. I've called Golden, a professor of history at Rutgers University and the author of *Message in a Bottle*, a book about the history of alcohol and pregnancy, to help me understand these once-common attitudes and practices. True to their perennial concern with weight gain, doctors tsk-tsked about the calories alcohol contributed to pregnant women's diets, says Golden. But there was little worry about the effect alcohol might have on the fetus. "Those in the medical establishment truly believed that alcohol couldn't hurt a pregnant woman or her fetus, and might help," she says.

This consensus was taught in hospitals and medical schools and codified in textbooks. "Alcohol, as such, is not injurious and need not be eliminated during pregnancy," read one obstetrics textbook published in 1953. "An occasional cocktail, highball, beer or ale, need not be restricted and may be most beneficial." Information intended for the general public often expressed even fewer reservations. In a 1964 book devoted to the significance of prenatal influences, the social scientist and popularizer Ashley Montagu dismissed the importance of one influence in particular. "It can be stated categorically that no matter how great the amounts of alcohol imbibed by the mother or the father, alcohol as such affects neither the germ cells nor the development of the fetus," he wrote. Women who wondered otherwise might receive a lecture from the experts. "No, smoking and alcoholic drinks have no effect on an unborn baby," chided a news bulletin distributed by a medical society in 1954. Calling such ideas "superstitions" and "old wives' tales," it concluded with a wag of the finger, "Listen to your doctor instead of sewing circle fantasy."

Doctors' confidence in the harmlessness of alcohol was based on their beliefs about the placenta. This organ, which implants itself in the uterus soon after conception to form a way station between woman and fetus, was thought to provide seamless protection from harmful substances. Medical historian Ann Dally traces this sanguine notion back to the attitudes of the late nineteenth century. "The Victorian tendency to put woman on a pedestal led to the idealization of the womb as well as of the woman," she writes, and to "a belief in the placenta as a perfect barrier against damaging influences." This conviction was still current in the 1950s, when Dally attended medical school; there she was taught that a toxin would affect the fetus only if it actually killed the mother. Pregnant women were not counseled about the dangers of medications or alcohol, Dally notes, and new drugs were not thoroughly tested for their safety during pregnancy.

And new drugs there were in abundance. The middle of the twentieth century was a golden age of pharmaceutical innovation, a time when serene sleep and steady nerves and a slim figure could be found inside the medicine cabinet. Pregnant women, too, were promised relief from all the complaints, small and large, of their condition: sleeplessness, morning sickness, miscarriage. The remedies were touted by advertisers in women's magazines, and by the publications' writers and editors themselves; an article about miscarriage in the November 1950 issue of *Woman's Home Companion* celebrated "the miracle drugs that have tumbled from the laboratories in such heartening profusion recently." The sales job worked: those who gave birth in the postwar years, writes one chronicler of the period, "were among the most medicated women in history." Between 1958 and 1965, according to one study, half of all new mothers took two to four pharmaceutical products while pregnant.

So buoyant was the belief in these chemical miracles that it took three successive tragedies to sink it: three stories of confident promises gone disastrously wrong. Learning about them is like following my fears about drugs during pregnancy to their long-buried source; I begin to understand why the subject stirs such profound anxiety in me and many others. The stories can be painfully difficult to read, especially now, but I find that I can't look away.

The first story begins in Germany on a summer day in 1961. Karl Schulte-Hillen, a lawyer and a new father, arrived at the children's clinic at the University of Hamburg bearing X-rays of his infant son. Jan had been born with shrunken arms and missing fingers; not long before his birth, Schulte-Hillen's sister had delivered a daughter with similar deformities. Schulte-Hillen had come to see the head of the clinic, pediatrician Widukind Lenz, in order to get some answers about the cause of these afflictions: Was it genetic? Was it something in the air or the water? Was it just a dreadful coincidence?

Lenz studied the X-rays, and agreed to help Schulte-Hillen find the source of his son's condition. The two men placed ads in newspapers, seeking children with similar defects. They drove all over the country in an old Volkswagen, showing people photographs of Jan and his cousin and asking if they knew of other such victims. (Don't be ashamed, Schulte-Hillen told them; this is a picture of my own child.) And Lenz began going through hundreds of thousands of Hamburg's birth records, stretching back decades. In twenty-five years, he found, there had been only one case of phocomelia, as Jan's truncated, flipper-like limbs were called; over the past thirteen months, there were fifty. Sure now that some exposure during pregnancy was to blame, Lenz started interviewing the mothers of affected children, asking them in detail about the daily routines they had followed while pregnant. On November 11, one of the mothers mentioned taking thalidomide.

Thalidomide was a remedy for sleeplessness and morning sickness, sold under reassuringly nonsensical names like Quietoplex and Softenon and Contergan. Its manufacturer, the German firm Chemie Gruenenthal, advertised the medication as exceptionally safe and completely free of side effects, though in truth the company had conducted little testing before bringing it to market. The following day, Lenz asked his interview subjects specifically about thalidomide, and four of the women replied that they'd taken the drug during pregnancy. It looked like Lenz had found the culprit—but Linde, Schulte-Hillen's wife, said she hadn't used the medication. Then she remembered that when she was four weeks pregnant, her father died, and her sister gave her a packet of thalidomide to help her sleep. She'd taken just one pill.

Lenz called a representative of Chemie Gruenenthal and told him that thalidomide must be immediately withdrawn. For good measure, he sent a letter to the company the next day: "Every month's delay in clarification," he warned, "means that fifty to one hundred horribly mutilated children will be born." The manufacturer balked, refusing to pull its profitable drug from the shelves. It acted only when a major German newspaper got hold of the story and ran it across its front page on November 25. Chemie Gruenenthal withdrew thalidomide that day, and the epidemic of malformed babies ended just about nine months later. It was too late, however, for the eight thousand children already deformed by exposure in utero. From a drug that was claimed to be free of side effects, thalidomide had become synonymous with the most awful side effects imaginable.

The second story begins six years later, on Independence Day in Massachusetts. Sixteen-year-old Shelley was enjoying the Fourth of July fireworks and softball games at her summer camp when she noticed blood on her underwear. Her mother took her to Massachusetts General Hospital in Boston, where gynecologist Howard Ulfelder diagnosed her with an extremely rare form of vaginal cancer. Called clear-cell adenocarcinoma, it was usually found only in postmenopausal women, but Shelley was the third young girl with the disease Ulfelder had seen in as many years. After examining her, Ulfelder came out of his office shaking his head. "There must be some explanation for this explosion," he told his nurse. The following month, he surgically removed Shelley's uterus and vagina, a step judged necessary to save her life.

At a checkup a year and a half later, Ulfelder was talking to Shelley's mother. I don't think I ever told you this, she said, but when I was pregnant with Shelley I took a drug called DES. Could that have anything to do with it? DES, or diethylstilbestrol, was a drug widely prescribed to pregnant women in the 1940s and 1950s to prevent miscarriage. Ulfelder said he couldn't imagine how the drug and the disease could be linked. But when he saw yet another teenager with the same rare cancer, he asked the girl's mother if she'd taken DES during pregnancy, and she said yes. "I thought, 'My god . . . ,'" he recalled. Ulfelder and two colleagues began seeking out other such cases. They found one in San Francisco, another in Mexico City, and yet another right in Boston. "You can stop your study," said a mother whose daughter had died of the cancer. "I took DES."

By 1971, the physicians had collected enough evidence to make an astonishing claim, which they put forth in the *New England Journal of Medicine*: a drug women took while they were pregnant was causing cancer in their teenage daughters. Shortly thereafter, the FDA issued an alert warning pregnant women not to take the drug. This was big news. Between 1947 and 1971, DES had been prescribed to around four million American women, and as many as eight million women worldwide. But DES had never been thoroughly tested for use during pregnancy, and it's doubtful that it was even effective in preventing miscarriage. Daughters exposed to DES in utero have forty times the risk of developing clear-cell adenocarcinoma (though the cancer remains rare even in this population), and they may experience other problems, such as infertility or pregnancy complications.

Most shocking of all was the way these risks came about: from a substance that crossed the placenta from mother to fetus. In just a few years, DES had gone from miracle pill to the world's first transplacental carcinogen.

The third and final story began in early February 1973, in an examining room at the University of Washington School of Medicine in Seattle. Kenneth Jones, a medical resident, and his mentor David Smith, a professor of pediatrics and an expert on birth defects, were called in to examine eight children. The young patients, who ranged in age from eleven weeks to almost four years, were all born to different mothers; two were white, three were black, and three were Native American. Yet four members of the group shared a kind of family resemblance: they were stunted in stature, with small heads, flattened faces, drooping eyelids, and smooth upper lips lacking the groove known as a philtrum. Each had been born to a woman who drank heavily during pregnancy.

The two doctors returned to Smith's office, where they pulled out his file of "unknowns," those cases Smith had encountered over the course of his career in which a birth defect could not be attributed to any definite cause. They quickly identified several patients whose characteristics matched those of the children in the examining room. Turning to these patients' medical charts, they saw that all had one thing in common: their mothers were chronic alcoholics. "At first, we really doubted ourselves," says Jones. "We thought, 'Could this be real?'" In fact, Jones was far from the first to wonder about the negative effects of drinking during pregnancy; such a relationship had been speculated upon for centuries. But Jones and his colleagues in Seattle were the first to define a formal diagnosis, which they did in an article in the medical journal the *Lancet* in June 1973.

"Fetal alcohol syndrome," as they named it, was character-ized by the distinctive features they had noticed in the examin-ing room, along with other problems like anomalies of the heart and the joints. Describing their subject group, which had come to include four additional patients, Jones and his coauthors wrote: "Eight unrelated children of three different ethnic groups, all raised in the fetal environment provided by an alcoholic mother, had a similar pattern of craniofacial, limb, and cardiovascular defects." Modestly, but firmly, they concluded, "We feel the data are sufficient to establish that maternal alcoholism can cause seri-ous aberrant fetal development."

At first, the new diagnosis encountered resistance: human beings have been drinking alcohol for millennia, some critics pointed out, and just now we start calling its effects a syndrome? Slowly, scientific and then social attitudes toward drinking and pregnancy began to change. In 1989, the federal government started requiring all alcoholic beverages sold in the United States to carry a warning label directed at pregnant women. Over the course of almost two decades, drinking during pregnancy had gone from a harmless and even beneficial practice to something quite different: the leading preventable cause of birth defects.

These three tragedies shocked us into the world we live in today, the only world I've known as a pregnant woman: one in which we are all too aware of the fetus's vulnerability, in which the placenta seems not an unbreachable barrier but the merest wisp. What *is* the placenta capable of blocking, I wonder, and what does it allow to pass through? To find out, I call Gideon Koren, professor of pediatrics and pharmacology at the University of Toronto and director of the Motherisk program at Toronto's Hospital for Sick Children. Motherisk runs the busiest hotline in the world for pregnant women with questions about exposures to drugs, chemicals, and diseases; its operators field hundreds of calls a week.

"During most of pregnancy, the placenta separating mother and fetus is only one cell thick," Koren tells me. "But it has an array of mechanisms to help it do its job of protecting the fetus." These subcellular tools, he explains, include tiny pumps that expel toxins before they can do any damage, immune agents that guard the placenta's perimeter, and placental enzymes that chemically break down intruding molecules.

This armamentarium does an impressive job of blocking bacteria from reaching the fetus, but it lets other substances sail right through. "The criteria that determine whether a molecule crosses the placenta include its size, its electrical charge, and its solubility," says Koren. "Not, notice, whether it is harmful or not." Particles that are small, that are neutrally charged, and that easily dissolve in fat will be waved past the placenta's layers of security, regardless of their potential toxicity.

Once they do cross the placenta, Koren continues, chemicals can affect fetuses more powerfully than adults, for several reasons: first, the fetus is so small that, pound for pound, it receives a higher relative dose of the chemical than would a person who's fully grown. Second, the fetus's detoxification and immune systems are still immature, unable to clear drugs and other chemicals from its system as effectively as the body of an adult. And third, the fetus is developing so rapidly that even a small disruption induced by a chemical can have far-reaching effects.

Science is still sifting through the fallout from its belated recognition of the risks posed by thalidomide, DES, and alcohol. It is evidence of the evolving state of our knowledge that the mechanisms by which these substances do their damage are not completely clear, even now. Thalidomide seems to wreak its devastation by inhibiting the growth of new blood vessels, preventing fetal limbs and other body parts from growing to their full size. DES appears to impersonate one of the body's own hormonal messengers, thereby disrupting the proper development of the female fetus's reproductive tract and making it more vulnerable to cancer later in life. And alcohol may trigger the death of cells in the fetus's growing brain, producing mental retardation and other neurocognitive deficits along with its distinctive effects on the anatomy.

The name for these substances, I now know, is *teratogen*: an agent that causes malformation of the developing embryo or fetus. The word is derived from the Greek word for monster, and indeed, historians have noted that many of the monsters of ancient Greek mythology, like the two-faced Janus and the one-

eyed Cyclops, appear to be based on types of congenital malformations. To this old and primal dread of a defective child I have
joined a new, more modern fear: of drugs and what they can do.
Before I began my reading this month, I knew little about thalidomide and DES, yet somehow the horror they inspired when
their malignity was revealed had seeped into my consciousness
and stayed there. My half-aware reaction, once I became pregnant, was to avoid anything like the drugs that led to those dimly
remembered disasters.

But this response is not universal, as I learn to my surprise.
Dozens of surveys have reached the same conclusion: women
swallow all manner of pills while pregnant. A 2005 study of 418
women, published in the *American Journal of Perinatology*, found
that more than three-quarters took at least one medication during
pregnancy; one-third took more than one, and 13.6 percent took
four or more. Several surveys have found that women take *more*
over-the-counter drugs when they're pregnant than when they're
not, and that pregnant women are taking more such drugs now
than in the recent past. The Slone Epidemiology Center Birth
Defects Study, which tracks 7,563 mothers and their children,
found that the use during pregnancy of five common nonpre-
scription drugs rose between 1976 and 2004. Pain relievers, cold
remedies, and allergy medications are the drugs most often consumed by pregnant women; a study of more than ten thousand
pregnant women, published in the *American Journal of Obstetrics
and Gynecology*, found that 65 percent took acetaminophen (the
active ingredient in Tylenol); 18 percent took ibuprofen (Advil or
Motrin) and 15 percent took pseudoephedrine (Sudafed).

Pregnant women are also prescribed drugs by their doctors.
A study of the records of eight HMOs, encompassing almost
a hundred thousand patients, found that 64 percent of women

were prescribed at least one drug during pregnancy; the Food and Drug Administration reports that while pregnant, American women take an average of three to five prescription drugs.

Most of these medications appear to have no effect on the fetus (though ibuprofen, for one, is not recommended for use during pregnancy, because of a very small risk of miscarriage or a fetal heart defect). My fear of taking so much as a Tylenol would seem to be overblown. But some scientists say we don't know nearly enough about how drugs work during pregnancy, including those commonly regarded as safe. "We're working without good, definitive information," says Martha Werler, senior epidemiologist at the Slone Epidemiology Center and professor of epidemiology at the Boston University School of Public Health. "Pregnant women should proceed with caution."

The FDA itself has acknowledged that its approach to labeling drugs for use by pregnant women is flawed. Its current system, devised in 1979, categorizes medications as A, B, C, D, or X, with A indicating little known risk to the fetus and X indicating a high risk. Doctors and patients have complained that the letter grades are confusing and inaccurate, and can't adequately accommodate emerging data about the drugs' safety. In spring 2008, the FDA proposed replacing the letters with more detailed information about risks to the fetus, and said it would require some drug manufacturers to fund studies on the use of their products by pregnant women. Such steps are long overdue. A 2000 study published in the medical journal the *Lancet* reported that 79 percent of pregnant women take at least one drug for which safety data are unavailable; even more troubling, the HMO study mentioned above found that almost one-half of pregnant women were prescribed drugs labeled C, D, or X.

Drugs don't come only in pill bottles, of course. The most

recent surveys by the U.S. Substance Abuse and Mental Health Services Administration indicate that 16.4 percent of pregnant women smoke cigarettes and 5.1 percent use an illicit drug. And, yes, some pregnant women report drinking alcohol: about one in five, again according to the federal government. (Given the social disapproval that attends drinking during pregnancy, women might be expected to hide their consumption of alcohol, but researchers have their ways. In one particularly sneaky study, conducted in Sweden on 103 pregnant women, scientists asked the subjects directly if they drank, and also analyzed samples of the women's hair and urine. Only about one in twelve admitted to drinking during pregnancy, but the chemical analyses found that one in four actually had.)

How bad is it to drink or smoke or take drugs during pregnancy? The answer that is now emerging is surprisingly complex. Smoking, it turns out, is very bad indeed. With each new study, it seems, another negative outcome is identified: smoking during pregnancy has been tied to miscarriage, stillbirth, preterm delivery, low birth weight, birth defects, and sudden infant death syndrome. More controversially, some studies have tied women's smoking during pregnancy to problems with their offspring's brain development and behavior in childhood, though it's not yet clear if these problems are caused by prenatal smoking or simply associated with it. We do know that maternal smoking affects the fetus in several ways: blood vessels in the woman and the fetus constrict, reducing the amount of oxygen and nutrition delivered to the fetus; carbon monoxide binds to the fetus's red blood cells, further reducing its oxygen levels; nicotine and hundreds of other chemicals cross into the fetal blood stream, potentially increasing its risk of cancer and other diseases. (Recent research suggests that pregnant smokers can still protect their fetuses from much of the harm of their habit—if they quit smoking before they reach week fifteen of their pregnancies. Scientists at the University of Auckland in New Zealand found that women who stopped smoking before the fifteen-week "deadline" had the same rate of premature delivery and underweight infants as pregnant women who did not smoke at all. Those who continued smoking, however, were three times more likely to give birth early, and twice as likely to have small babies, as nonsmokers.)

Even as evidence of the harm caused by smoking mounts ever higher, the effects of in utero exposure to crack cocaine seem to be not as devastating as once feared. When the crack cocaine epidemic whirled through America's inner cities in the 1980s, there were ominous warnings about a generation of irreparably damaged "crack babies." The Maternal Lifestyle Study, which has followed more than a thousand such children since infancy, has found that the worst-case scenarios have not come to pass. An evaluation of the babies' neurodevelopment at one month of age concluded that while effects of cocaine exposure were detectable, they were subtle; followups at ages three, five, and seven years found that while high prenatal cocaine exposure was associated with behavior problems, these effects were not as significant as those produced by their mothers' use of alcohol and tobacco.

Heavy drinking, surprisingly enough, appears to be worse for the fetus than the use of crack cocaine. But what about more moderate imbibing—say, a glass of wine a couple of times a week? Here the research veers first one way, then the other; a reader venturing into this literature is liable to get whiplash. One set of findings suggests that moderate drinking during pregnancy is not harmful—and, at least according to one peer-reviewed study, might even be beneficial. In a finding that lit up the email in-boxes of my pregnant friends, a 2008 study of 12,500 three-year-olds found that the female children of women who were light drinkers during pregnancy (defined as two or fewer drinks a week) were less likely to have emotional problems than the girls of women who did not drink at all. Male children of light drinkers were less likely to be hyperactive or to have conduct disorders, and scored higher on tests of cognitive ability than the boys of abstainers.

The study's authors, from University College London, note that light drinkers tend to be better educated and from higher-income households than abstainers, and that these socioeconomic characteristics, rather than maternal alcohol consumption itself, could explain some of the results. (That was another surprise to me: as this and other studies have found, women who drink moderately during pregnancy are older, better-educated, more affluent, and have better prenatal care than women who don't drink at all.) But, added lead author Yvonne Kelly, "it may also be that light-drinking mothers tend to be more relaxed themselves, and this contributes to better behavioral and cognitive outcomes in their children."

"Does this mean I can hit the bar on my way home?" emailed one pregnant friend facetiously (I think). Before you do, I told her, you should take a look at a countervailing body of research indicating that even moderate drinking during pregnancy is associated with behavior, learning, and attention problems in offspring. Studies in animals also suggest that exposure to even small amounts of alcohol can teach offspring to like its taste, making them more likely to drink themselves when mature.

The stubborn inconclusiveness of the evidence regarding the effects of moderate drinking during pregnancy has led doctors and public health officials to make rather cryptic and evasive statements on the subject—in fact, the *same* cryptic and evasive statement, as if they're all reading from the same cautious script.

"There is no known amount of alcohol that is safe to drink during pregnancy," recites my obstetrician, and the authors of the pregnancy guides on my shelf, and the spokespeople from the March of Dimes and the National Institutes of Health and the Centers for Disease Control. Despite my own aversion to the notion of drinking while pregnant, my contrarian impulses flare in the face of this rehearsed uniformity. What about all the women of a generation or two ago, I ask myself, who drank during pregnancy and produced perfectly healthy babies?

I decide to put the question to Ken Jones, the doctor who defined fetal alcohol syndrome more than thirty-five years ago, and who is now chief of the teratology division in the department of pediatrics at the University of California, San Diego. When I reach Jones, a genial-sounding man, on the phone, my query comes out in an unexpectedly personal form.

"My mother drank when she was pregnant with me," I tell him, "and look, I'm fine." I hear a slight quaver of anxiety in my voice. Jones's reaction is not reassuring: he laughs.

"My first response to that would be, 'Are you *sure* you're fine?'" he says. "What could you have been if she didn't drink?"

"But seriously," he continues, "there's a lot we don't know about the effects of alcohol during pregnancy." Pregnant women, and their fetuses, may vary in their vulnerability to alcohol, he explains; just as some adults can throw back a few beers every night while others are staggered by half a glass, individual women and fetuses may metabolize alcohol at different rates. (This may hold true for other substances as well: a 2002 study found that pregnant women who smoke are more likely to have a premature or low-birthweight baby if they have either of two common genetic traits; these traits influence the body's ability to dispose of certain chemicals.) Alcohol consumption during pregnancy may also combine with other factors, such as nutrition or stress, to affect some fetuses differently than others. Even among women who drink heavily during pregnancy, Jones points out, not all will deliver a baby with fetal alcohol syndrome. In fact, one study reports, while 71 percent of the babies of poor alcoholic women had fetal alcohol syndrome, that was true of only 4.5 percent of the infants of more affluent alcoholics, probably because of their more nutritious diets.

"We just don't know when drinking might be safe, and for whom," Jones concludes. "We can't be sure, so we tell women: 'Don't drink during pregnancy.' Seems like a small price to pay to me."

Easy for him to say. But, weighing the evidence, I decide to stick with my teetotalling ways. And I'll continue to stay away from most medications, now that I've learned how little we still know about their effects on the fetus. But what about all the chemicals that I *can't* avoid?

When you become a parent, you leave behind a world constructed of solid wood and clear glass and soft fabric to enter a universe made of plastic: wipeable, unbreakable, primary-colored plastic. Each morning, it seems, I trip over plastic cars on my way to the kitchen; there I serve Teddy his toast on a plastic plate, his milk in a plastic cup; after he's finished, he plays with his plastic toys while I place his lunch in plastic containers. Like the miracle drugs of the last midcentury, plastic was supposed to make our lives easier and better, improving on what nature was able to accomplish; in the 1950s, scientists working for General Electric dreamed of creating "synthetic man," a human made of plastic parts. But as with thalidomide and DES, somewhere along the line plastic turned on us, becoming the threat from which we need protection.

My introduction to the hazards of plastic came in the spring of 2008, a few months before I got pregnant. One morning in mid-April, my email in-box began to fill with forwarded messages from other parents, the twenty-first century version of the phone tree, now buzzing with alarm. The emails had subject lines like "Please Read—Very Important," and "Scary!" A federal agency called the National Toxicology Program had released a preliminary report on bisphenol A, a synthetic chemical found in many plastic products, including baby bottles and food and water containers. Following a review of hundreds of studies performed on laboratory animals, the report expressed "some concern" that fetuses, babies, and children could be harmed by BPA. "These studies only provide limited evidence for adverse effects on development, and more research is needed to better understand their implications for human health," it read. "However, because these effects in animals occur at bisphenol A exposure levels similar to those experienced by humans, the possibility that bisphenol A may alter human development cannot be dismissed."

The report's cautious language did little to allay the anxieties of my email correspondents. Their concerns seemed confirmed a few days later, when word came that Canada's health department had officially designated BPA a "dangerous substance." Overnight, it seemed, BPA fever spread through my Upper West Side neighborhood. A previously obscure industrial chemical was now the talk of the playground and the supermarket; former liberal arts majors started sounding like chemistry PhDs, and their children began toting their apple juice in shiny metal thermoses instead of plastic sippy cups. Something similar was happening on a national scale. Consumers—many of them worried parents—demanded BPA-free goods, and companies scrambled to oblige. Within a few weeks the baby-bottle manufacturer Playtex and the sport-bottle maker Nalgene announced that they were removing BPA from their products, and Wal-Mart, CVS, and Toys "R" Us promised to begin removing BPA-containing items from their stores.

This all happened so quickly that it made me wonder if we'd been suspicious of plastic all along, had known it was too good to be true. Plastic is the very definition of unnatural: it doesn't splinter like wood or crack like glass, and indeed the property contributed by bisphenol A, the reason it is used so widely, is that it makes plastic shatterproof. There had to be a hidden cost to such convenience, and now we were learning what it was.

In fact, an early clue to the potential risks posed by plastic was discovered entirely by accident. In August of 1998, Washington State University geneticist Patricia Hunt found herself baffled by an unexpected development in her lab. Hunt was using experiments on lab mice to explore why the eggs of older women are more likely to produce children with chromosomal abnormalities such as Down syndrome. The research was coming along well when suddenly the eggs of her control animals—the normal, healthy, young ones—"went goofy," she told me, spiking from 1 percent abnormal to 40 percent abnormal. Mystified, Hunt checked every variable she could think of: the food fed to the mice, the air circulating in her lab, the medium used to culture the mouse eggs.

The answer, when she found it, was startling. A temporary lab worker had mistakenly washed the mice's plastic cages and water bottles with a harsh detergent, damaging the plastic and releasing small amounts of bisphenol A. At first, Hunt found it hard to believe that BPA could be the cause of the mice's abnormal eggs. She checked and checked again, exposing a new set of animals to the damaged cages and bottles, even feeding bisphenol A directly to yet another group of mice. All the results pointed to BPA as the culprit. Still, she says, "I sat on the data for several years, because I wanted to make sure that I got everything right. I realized that what I would be saying was that low doses of this chemical could be dangerous. And that's pretty serious stuff."

Her finding was serious, and controversial, because it contradicted an age-old tenet of toxicology: the dose makes the poison. Scientists' understanding of the effects of chemicals—and, not incidentally, the nation's regulatory system—has long been based on the notion that the bigger the exposure, the greater the harm, and that even known toxins are harmless in small amounts. But BPA "doesn't play by those rules," says Hunt. She and many other researchers say that BPA is an endocrine disruptor, one of a group of chemicals (which also includes the drug DES) that mimic the action of our bodies' own hormones—and, indeed, were often invented to do just that.

DES, for example, was first synthesized in 1938 by a British scientist, Charles Edward Dodds, who had set out to develop an artificial version of the female hormone estrogen that could be easily and cheaply produced. Diethylstilbestrol was three times more powerful than naturally occurring estrogen—so potent, in fact, that male scientists who inhaled the substance while working in Dodds's lab began to grow breasts. Bisphenol A also first attracted notice in the 1930s as a synthetic estrogen. When plastics began to be mass-produced some two decades later, manufacturers added various chemicals, including BPA, to improve the plastics' usefulness: to make them soft and pliable, to make them hold a scent, or (in BPA's case) to make them hard and clear.

Today, six billion pounds of BPA are generated each year for use in plastic.

Hormonelike chemicals like DES, and likely BPA, are not toxic in the familiar fashion of, say, lead or asbestos; rather, they wreak havoc by interfering with the body's natural hormonal signaling system. They may be capable of disturbing development even in small amounts—indeed, they may be *especially* disruptive in small amounts. At high doses, our bodies recognize them as foreign, and protective mechanisms kick in; at low doses, our bodies are lulled into a dangerous complacence. Given that development is never more rapid or more consequential than during gestation, fetuses are especially vulnerable to the action of endocrine disruptors. It may be necessary to rewrite the old maxim: it's the *timing*, not the dose, that makes the poison.

Following the discovery of the abnormal mouse eggs in her lab, Hunt decided to make an investigation of the effects of BPA the focus of her research. Over the next few years, she and a handful of other scientists, like Frederick vom Saal of the University of Missouri, Columbia, and Shana Swann of the University of Rochester, piled up a small mountain of animal research implicating BPA in hormonal and reproductive abnormalities and increased risk of diseases such as cancer and diabetes. Their findings quietly and steadily accumulated until the National Toxicology Program's report brought them forward in 2008, setting parental alarm bells ringing all over the country.

Many scientists say that the weight of the evidence justifies such concern; others, including representatives from the plastic and chemical industries and some government officials, say that parents' fears are overblown, and that there is no proof that BPA is harmful to humans. Amid all the claims and counterclaims, one story sticks in my mind. I hear it again one evening as I'm listening to a tape of my interview with Patricia Hunt. "After I figured out the connection between my mice's abnormal eggs and their exposure to bisphenol A, I sat down in my study to read everything I could find about BPA," Hunt says in her soft voice. "After half an hour, I was so agitated that I stood up, went to my kitchen, and threw all my plastic containers in the trash."

I click off the tape and sit for a moment, lost in thought. Then I get up and go into my own kitchen. I open a cabinet spilling over with plastic containers: Teddy's plates and bowls and cups, my own Tupperware and water bottles. I inspect the bottom of each one: anything labeled with the recycling code three, six, or seven goes into the trash. (In my research on BPA, I encountered a mnemonic device with the lilt of a nursery rhyme: "Four, five, one, and two/All the rest are bad for you.") I check the rest for signs of wear: anything with a roughened surface gets pitched into the garbage, too. Following them into the trash can are the baby bottles I used when Teddy was an infant; when this baby is born, I'll buy the BPA-free kind. I regard what's left with a wary eye. Some things are just too convenient to get rid of: Teddy's preschool, for example, requires reusable containers in his lunchbox. They'll remain in my kitchen on probation—but they won't be going in the microwave or the dishwasher, since heat, like hard use, can release BPA.

Returning to my home office, I feel better, even though I know from my reading that my kitchen purge is only one small step in reducing my family's exposure to bisphenol A. BPA is all around us, in coffee makers and food processors, bike helmets and car parts, water filters and dental sealants. It's in the computer under my fingers and in the eyeglasses perched on my nose.

And BPA is not the only "plasticizer" to come under scrutiny of late. Phthalates are synthetic chemicals added to plastics to make them soft and pliable, or to products like perfumes and cosmetics to help them hold their scent. Studies in animals have found that exposure to phthalates during gestation disrupts the proper development of male offspring, resulting in incomplete "masculinization." In humans, phthalate exposure has been associated with a higher risk of an abnormality of the penis called a hypospadia, in which the opening of the urethra is located in the wrong place. A 2008 study published in the journal *Environmental Health Perspectives* found that hairdressers and beauticians frequently exposed during pregnancy to hairspray (which often contains phthalates) were 2.3 times more likely than other women to give birth to sons with hypospadias.

Phthalates, like BPA, are found in products we use every day—in shower curtains and car upholstery, lotions and shampoos—and their presence is rarely noted on labels. (In the European Union, phthalates are now banned from cosmetic products; in the United States, they are banned from children's toys.) Like a drink with dinner or a pill for a headache, plastic has become deeply, perhaps inextricably, embedded in our everyday lives.

I turn the recording of my interview with Patricia Hunt back on; she's talking about the need to regulate potentially dangerous chemicals like BPA, and about the risks of waiting to take action until such chemicals have already caused harm. She speaks from her own experience. "I'm a DES daughter," Hunt says. "My mother took the drug when she was pregnant with me." Though Hunt has been spared the vaginal cancer that made DES infamous, she has endured a bout with breast cancer, which is two-and-a-half times more common among over-forty women prenatally exposed to DES than among women not exposed. Noting that bisphenol

A was once considered for use as a synthetic estrogen, only to be abandoned in favor of the more potent DES, Hunt says she finds it uncanny that both of these chemicals should have found their way into her life. "I think of my DES exposure and the accident with BPA in my lab as two lightning strikes," she says. The DES saga in which she has a part holds lessons, she believes, for the current debate over BPA. "People say to me, 'But your research is with mice. Are your findings relevant to humans?'" says Hunt. "I tell them, 'With DES, we ran the experiment on humans first. Do we really want to do that again?'"

Walking Teddy to school the next morning, plastic-filled lunch-box in hand, I take a wary look around. Everything looks toxic to me today: the idling trucks outside the supermarket, belching exhaust; the lurid green pools of antifreeze, left at the curb by departing cars; the smoke curling from loiterers' cigarettes, held at the height of a preschooler's nose. The research I've been doing only deepens this impression: every day, it seems, brings a new finding about the danger to fetuses posed by environmental exposures. Cell phones emit radiation; dental fillings contain mercury; pajamas are treated with flame-retardant PCBs. (That one really gets me: dangerous *pajamas*?)

As Teddy and I make our way up Broadway, my thoughts turn to a study I read just last night. Beginning in 1998, more than five hundred pregnant women fanned out across the South Bronx and Upper Manhattan, a few blocks from where we're walking now. All the women had on identical black backpacks, which they wore every waking moment for two days: in their homes, at work, around their neighborhoods. At night they hung the bags on a chair next to their beds. Inside each backpack was an air monitor, continuously measuring levels of a type of pollutant called poly-cyclic aromatic hydrocarbons, or PAHs. (PAHs come from vehicle exhaust; they are also present in the fumes released by cigarettes and factory smokestacks.)

A check of the monitors revealed that 100 percent of the women were exposed to PAHs during their pregnancies. After their babies were born, analyses of umbilical cord blood from sixty of the infants showed that 40 percent had subtle DNA damage from PAHs—damage that has been linked to an increased risk of cancer. Further analysis found that those exposed prenatally to high levels of PAHs were more than twice as likely to be cognitively delayed at age three, scoring lower on an assessment that predicts poor performance in school; at age five, these children scored lower on IQ tests than children who received less exposure to PAHs in the womb. Researchers expect to find more effects as the children grow older.

To my mental landscape of looming dangers, add: the air. The psychic shift I've undergone since I started reading about chemical exposures and pregnancy reminds me of the way the world transformed itself once Teddy began reaching and crawling. Suddenly, my plane of vision was full of steps steep enough to tumble down, objects tiny enough to choke on. How much of my new consciousness of chemicals is unfounded alarm, I wonder, and how much necessary parental protectiveness?

Frederica Perera, the director of the Center for Children's Environmental Health at Columbia University, is the lead investigator of the backpack study; when I return home from school drop-off, I give her a call. Perera tells me that her interest in the effects of pollution on fetuses began more than thirty years ago, when she was doing research on environmental exposures and cancer in adults.

"At the time, I was looking for control subjects to compare to the adults in my study—individuals who would be completely untouched by pollution," she says. She hit on the idea of using babies just out of the womb as her controls. "So, I got some samples of cord blood and placental tissue and sent it off to the lab, and back came the results. I took one look, and then called the lab and told them, 'There must be a mistake, somebody goofed— rerun the analyses, please.' They called back and said, 'Nope, those results are correct. We redid the tests and got the same numbers.' I was shocked, because these samples I thought would be pristine already had evidence of contamination."

She soon switched the focus of her research from adults to fetuses; since then, research by Perera and others has tied exposure to traffic-related air pollution during pregnancy to a host of adverse birth outcomes, including premature delivery, low birthweight, and heart malformations. A 2002 study led by UCLA epidemiology professor Beate Ritz, for example, found that the babies of women living in areas of Los Angeles with severe air pollution have three times the incidence of heart malformations and valve defects. A 2007 study, also by Ritz, found that pregnant women living in areas with high carbon monoxide or fine particle levels have a risk of preterm birth that is 10 to 25 percent higher than those who live in neighborhoods with cleaner air.

The pollutants under scrutiny include not just PAHs, but also carbon monoxide, nitrogen dioxide, ozone, and particulate matter. These substances may do their damage by undermining the pregnant woman's health, by impeding the proper functioning of the placenta, or by entering the fetus's own bloodstream. Pollutants may even alter gene expression in the developing fetus; Perera and her colleagues at Columbia are now exploring whether children exposed prenatally to high levels of traffic are at greater risk of developing asthma because of epigenetic changes they acquired in the womb. Investigations like these have prompted a shift in the object of scientists' concern about the effects of pollution. "We used to worry about old people and asthma patients," Perera tells me. "Now we worry about fetuses."

One of the tools these researchers employ to gauge the impact of pollution is biomonitoring, a technology that allows them to measure the levels of environmental chemicals found in bodily fluids and tissues. Biomonitoring studies are beginning to show just how many of these chemicals we carry within us, constituting what is known as our "body burden." Surveys by scientists at the Centers for Disease Control have found bisphenol A, phthalates, and PAHs, as well as traditional toxins like mercury and lead, in the blood or urine of most people tested.

There's no evading the chemicals that permeate our planet, it seems: the streams of affluent suburbs burble with discarded Prozac and Paxil, and traces of flame retardant turn up in the bodies of polar bears in the wild. The scientists' dream of "synthetic man" has in a sense come true; we're all partly plastic now. I'd like to believe, as Frederica Perera once did, that the womb is the last unspoiled place, but man has left his tracks even here. The umbilical cords of newborn babies contain an average of two hundred industrial chemicals. As Perera's backpack study showed, we carry our fetuses with us into the world; they are out on the traffic-clogged street, present at the smoke-filled party, recipients of the drinks we sip and the drugs we swallow. They shoulder part of our body burden before they are even born.

What do the presence of these chemicals in our bodies mean for our health, and for the health of our fetuses? The candid answer is that we don't know. Skeptics point out that just because a chemical is in us doesn't mean it's doing us harm. Indeed, our nation's regulatory system has long operated on the assumption that unless a chemical is proven toxic, it's all right to have around. Some 80 percent of major chemicals used in industry have never been evaluated for their effects on early development.

But some scientists and public health advocates are proposing that we turn that assumption on its head: chemicals should be proven safe first, before we allow them into our lives (and, inevitably, into our bodies). This idea is known as the "precautionary principle," and one of its most thoughtful proponents is Philippe Grandjean, chair of environmental medicine at the University of Southern Denmark and adjunct professor of environmental health at Harvard University. "The precautionary principle shifts the burden of proof," Grandjean explains. "The producer of a chemical would have to demonstrate the absence of risk, rather than the public authorities having to demonstrate harm."

Early in his career, Grandjean studied the effects of childhood exposure to lead; later, he was one of the first to sound the alarm about the effects of prenatal exposure to mercury. In both cases, he says, policymakers waited far too long to take action, allowing millions to be adversely affected. Now, he tells me, he sees the same thing poised to happen again with bisphenol A and other environmental chemicals. "Whenever we assess the risk posed by a chemical, we have to consider its effects on individuals at the most vulnerable stages of life, including the fetal stage," he says.

As individuals, we can implement something like the precautionary principle in our own lives. We can remove suspect chemicals from our homes; we can check labels and choose our purchases more carefully in stores. But even the most committed pregnant woman can do only so much on her own. Real change must come collectively, in the form of more thorough testing, more informative labeling, and more restrictive regulation. And that change must come soon, adds Grandjean: "We need to learn from the mistakes of the past, or we'll end up with the same costly consequences." As he speaks I find that I'm vigorously nodding my head; when I hang up the phone, I feel the not-unpleasant flush of righteous indignation.

And then, toward the end of my fourth month—just as I begin to exhale an anxiety I hadn't fully acknowledged, the fear that an early pregnancy won't last—I get a call from a friend whose due date is a few weeks after mine. We had been delighted to discover that we were pregnant at the same time, had laughed as we imagined pushing our carriages side by side down the sidewalk.

"I lost the baby," she says quietly.

In the wave of emotions that sweeps me—sadness, shock—there's another: shame. When reading about DES, I realize, I had mentally passed judgment not only on the doctors who casually dispensed the drug, but also on the women who obediently swallowed it. Why weren't they more careful? I wondered. Why didn't they ask more questions? The pain in my friend's voice is a rebuke to my high-handed attitude. If a drug that prevents miscarriage were available, wouldn't my friend—the survivor of a previous pregnancy loss—have taken it and hoped for the best? Wouldn't the same be true of the friend who experienced life-threatening preeclampsia, or the one who suffered debilitating morning sickness, or the one who went into labor two months early? In each case, I realized, their pharmaceutical options were severely limited: there were few effective drugs available, or else little was known about the drugs' safety during pregnancy.

In fact, some researchers have been raising alarms about a so-called drug drought in maternal medicine: a critical lack of effective treatments for pregnant women. Spooked by night-mares like thalidomide and DES, pharmaceutical companies have avoided developing or testing drugs for pregnant women, and industry and government drug trials exclude pregnant women as a matter of course. "Only one new class of drug has been licensed for obstetric conditions in the last twenty years, and the situation is set to worsen," with no new class of drug for use in pregnancy now in clinical trials, reports a 2008 article in the journal *PLoS Medicine*. Only seventeen of the thirty-seven thousand drugs under development worldwide since 1981, the study finds, were intended for use by pregnant women. That's less than 3 percent of the number of drugs developed in that period to treat heart disease, and less than half the number developed for Lou Gehrig's disease, of which there are fewer than six thousand new cases a year in the United States (as compared to four million births). If pregnant women like me live in a premodern, prepharmaceutical world, I realize, it's not entirely by our choice.

And yet I find the notion disorienting to contemplate: there are not *enough* drugs for women to take during pregnancy. A pill that prevented preeclampsia or arrested premature labor would be a godsend for many women, of course. Many others must take medication to manage preexisting conditions during pregnancy. Still, I can't quite overcome my aversion to the idea of taking a drug while pregnant—until I remember that I already do. There is a pill that I, along with millions of other pregnant women, swallow without qualm every day: folic acid. One morning, I shake one into my palm and stop. What is this, I wonder, and why is it recommended so enthusiastically to pregnant women? It turns out that there's a story behind this round white tablet, just as there were stories behind thalidomide and DES.

Back in 1983, British scientist and doctor Nicholas Wald set out to help mothers who had delivered a baby with spina bifida, an often-devastating birth defect in which the spine and its protective covering fail to completely develop. These women were ten times more likely to have another child with spina bifida. No one knew how to prevent it; mothers of children with the condition were often told that its cause was purely genetic. "Many of these women were terrified to have another baby, fearing that their next child would also be affected," Wald tells me when I reach him in London, where he is now the director of the Wolfson Institute of Preventative Medicine.

Wald and others had noticed that spina bifida was more common among the offspring of poor women with inadequate diets. Could some nutritional deficiency, he wondered, be causing this malformation in the womb? Wald set out to test this possibility as the principal investigator of the Medical Research Council Vitamin Study, involving thirty-three hospitals in seven countries. It enrolled 1,817 mothers of children with spina bifida or other, similar deformities, collectively known as neural tube defects. The women were divided into four groups: some received supplements of folic acid, a B vitamin; some received a mixture of other vitamins; some received both, and some neither. The women who took the folic acid were 72 percent less likely than those who did not to have a second child with a neural tube defect. "When we saw those results, we stopped the study early," says Wald.

It was now clear to him that every woman who is pregnant, or even contemplating pregnancy, should be receiving this drug. Wald published his findings in the *Lancet* in 1991; the following year, the U.S. Public Health Service issued a recommendation that all women of childbearing age take a daily folic acid supplement. In 1998, the government went a step further. It added folic acid to the nation's food supply, fortifying breads, cereals, and other grain products with the vitamin. Since then, the rate of neural tube defects in this country has dropped by more than a third.

A simple pill that prevents birth defects: folic acid actually produces the kind of miracles once promised to pregnant women by the pharmaceutical companies. Because of Wald's research, thousands of babies who would have suffered from a crippling disability were instead born healthy and whole. Imagine the drugs of the future that could do the same for other problems originating during pregnancy: the hundreds of thousands of premature babies born in the United States each year; the quarter-million pregnant women who suffer from preeclampsia; the 3 to 5 percent of infants born with birth defects, most of unknown origin.

The healthy pregnancy of the future may well lie in restricting exposure to dangerous chemicals, and taking full advantage of beneficial ones. The trick, as always, will be telling the difference.

FIVE MONTHS

I want a boy. Is it all right to say that out loud?

I want a boy because I already know about girls. Growing up I had one sister, and thirty female classmates at my all-girls' school; childhood was a clutter of hair ribbons and Barbie shoes, adolescence a mist of Secret deodorant and Love's Baby Soft perfume. I knew almost nothing of boys until I gave birth to one, which may be why they captivate me now.

It's an unexpectedly mild November morning on the rooftop playground of Teddy's preschool. I'm sitting with a gaggle of other parents, watching the Darwinian contests play out like a nature film about lions and antelope on the savanna. "Maya, we don't pinch our friends!" the mother seated next to me calls out. "She's *not my friend*," Maya hisses over her shoulder. The girls' urgent confidences and casual insults come back to me like a decades-old echo. But the boys are exotic, of almost zoological interest: Look how one pushes another to the ground, then offers a friendly hand up! Watch how they shoulder into one another, butting and shoving like young animals everywhere! In the middle of the scrum is my own boy, his face open, laughing. I want another one.

But along with my wish comes a host of misgivings. I'm uncomfortably aware that my preference, personal though it feels, is one shared by cultures throughout time and all over the world: from the ancient Greeks, who believed that a proper pregnancy produced a boy while a girl was evidence of error, to present-day China and India, where millions of female fetuses have been aborted in the drive to have a son. I'm disturbed to find myself aligned, even unintentionally, with these twisted attitudes and practices. And, too, my wish feels dangerous in its desire to meddle with the natural order of things; it seems like hubris, ripe for retribution. Will a bolt of lightning strike this rooftop, smiting me down because I dared to interfere with the divine plan?

In fact, the man who first documented the ratio of male births to female births regarded their roughly equal proportion as proof of the existence of God. John Arbuthnot was physician to the Queen of England, consort of Jonathan Swift and Alexander Pope, and a pioneering statistician who in 1710 published "An Argument for Divine Providence, Taken from the Constant Regularity Observed in the Births of Both Sexes." Using the bills of mortality—weekly tallies of births and deaths—compiled by London parish clerks between the years 1629 to 1710, Arbuthnot showed that the ratio of male babies to female babies was remarkably consistent, and declared that this was not the result of chance. Slightly more males than females were born each year, he noted, an excess that accommodated "the external accidents to which males are subject."

Such careful calibration ensured pairs as neat as those on Noah's ark: one man for every woman. "Among innumerable Footsteps of Divine Providence to be found in the Works of Nature, there is a very remarkable one in the exact Ballance that is maintained between the Numbers of Men and Women," Arbuthnot wrote approvingly, "for by this means it is provided, that the Species may never fail, nor perish, since every Male may have its Female, and of proportional age." To his mind, this balance was proof of a wise creator, and evidence that polygamy and celibacy are "contrary to the Law of Nature and Justice." Protestant ministers saw it that way, too, hailing Arbuthnot's findings from their pulpits.

"An Argument for Divine Providence" was also a scientific landmark, the first known use of inferential statistics. Remarkably, a branch of mathematics was invented in order to examine how many boys and girls are born. It's an early example of a persistent theme: the gender of babies always comes freighted with meaning, of a worldly and even of a metaphysical kind.

But even if, as Arbuthnot insisted, the sex ratio is divinely ordained, we still want advance notice of the plans God (or nature, or chance) has for our children. The desire to know whether a baby will be a girl or a boy is an old and apparently universal one. When Egyptologists translated the three thousand-year-old scroll known as the Berlin Papyrus, they found themselves spelling out directions for a prenatal sex-identification test. "You shall put wheat and barley into purses of cloth, the women shall pass her water on it every day (it being mixed with dates and sand)," directed the scroll. "If both sprout, she will give birth. If the wheat sprouts, she will give birth to a boy. If the barley sprouts, she will give birth to a girl. If they do not sprout, she will not give birth at all." (Incidentally, this isn't a bad test of pregnancy: a twentieth-century re-do found that 70 percent of the time, the urine of pregnant women did indeed promote faster growth of seeds, perhaps because of the hormones it contains. But there's no evidence that the procedure tells us whether the fetus is male or female.)

Other alleged indicators of fetal sex, gathered from my reading, are similarly ingenious: Women carrying boys are in good spirits, while women bearing girls have volatile moods. Dreams about knives or clubs mean a boy, while dreams about spring or parties signify a girl. Morning sickness means it's a girl; a big appetite, a boy. A woman favors her right side when she is pregnant with a boy; her left side, with a girl. A fetus who is quiescent is female; a fetus who moves early and vigorously is male. The future mother of a male child has rosy cheeks, while her counterpart with a female child is pale.

There's something about the opaque orb of the pregnant belly that invites such speculation: it's a silent oracle, a blank crystal ball. The sex of the fetus within becomes the hook on which we hang our multiplying questions about what the future child will be like. I'm surely no exception: I feel sheepish when I realize that I've been applying each of these tests to myself—and then disappointed when the results turn out to be inconclusive. Pale skin, left-handed: maybe it's a girl. Big appetite, no morning sickness: could be a boy. It makes me wonder. Why do I—why do we—want to know so urgently? At the end of nine months, the answer will be clear. Why do we need to know *now*?

Sometimes, great consequences do hang on the matter of male or female: a rich inheritance, a royal succession. Rarely have the stakes been higher than they were for Anne Boleyn, pregnant with King Henry's child in the spring and summer of 1533. A baby boy would banish the questions that still lingered around her recent marriage to Henry, which followed his shocking divorce from his first wife, Catherine of Aragon. In eager anticipation of a male heir, the king had already picked out the names Henry and Edward, and planned tournaments and concerts, jousts and pageants, to herald the birth of his long-awaited son. His confidence was bolstered by the predictions of court astrologers, physicians and prophets, who all said that the baby would be a boy.

The seers were wrong, of course. At three o'clock on Sunday, September seventh, Anne gave birth to a daughter, Elizabeth. The festivities were called off, and the birth announcements, already lettered, were amended to turn "prince" into "princess." When Anne's second baby, a male, was stillborn, Henry divorced her, and had her beheaded to boot. (The erring astrologers and prophets escaped punishment.)

Until word of the newborn Elizabeth came down from Anne's chambers that September afternoon, nothing at all was known about the baby. It's hard to overstate how little firm information we had about the fetus until recent times—its sex, its size, its health. For early knowledge of the very existence of the fetus, physicians since the days of Hippocrates relied on the word of the woman herself. This made quickening, or the first fetal movements perceived by the woman, a signal event. Samuel Pepys, the exuberant and indefatigable chronicler of seventeenth-century London life, recorded just such an event in his famous diary. On New Year's Day of 1663, he wrote of the mistress of another king of England, Charles II. Barbara Palmer, Lady Castlemaine, had "quickened at my Lord Gerard's at dinner, and cried out that she was undone," Pepys reported breathlessly, "and all the lords and men were fain to quit the room, and women called to help her." In an instant, Lady Castlemaine was transformed from a mere paramour to the future mother of the king's child; the tiniest flutter of a fetus inside, apparent only to her, was enough to alter her status (and to clear the room of the menfolk).

In the centuries that followed, quickening took on additional significance as a key moral and legal benchmark. In the United States as well as in Britain, quickening came to mark the official beginning of the fetus's existence. Life, declared the legal authority *Blackstone's*, "begins in the contemplation of law as soon as an infant is able to stir in the mother's womb." Until at least the mid-nineteenth century, abortion was not regarded as a crime if it occurred before quickening. And this information could come only from the woman: what we knew of the fetus, we knew from the inside.

How little we knew from the outside is made clear to me on a morning I spend in Columbia University's Butler Library. Under its vaulted ceilings, in the light filtering down from its high windows, I page through several centuries of artists' imaginings of the fetus. A drawing from sixteenth-century France shows a tiny, forlorn figure, lost in the vast sea of his mother's uterus, a length of umbilical cord his only link. One from seventeenth-century Italy depicts a pair of twins standing upright in the womb, their arms thrown around each others' waists as if out for a stroll. In a sketch from eighteenth-century Germany, the fetus appears as a muscled strongman, about to burst the walls of the womb. Perhaps strangest of all is a diagram by the nineteenth-century German biologist Ernst Haeckel. It looks to show a fetus performing a zoological costume change—from minnow to crawfish to chick to possum, emerging triumphantly as a gill-less, wingless human. The drawing illustrates Haeckel's influential theory of "recapitulation," which held that the fetus passes through all the stages of humankind's evolution in the course of pregnancy, an eon of change squeezed into nine months.

These images are so alien, so unsettlingly odd, that two exceptions stop me in my tracks. The first are photographs of fetal mannequins, made by Italian artisans in the eighteenth century. Giovanni Antonio Galli, a professor of obstetrics at the University of Bologna, used the models to help teach midwives how to assist at a birth. Sculpted from wax and clay, they are startingly lifelike, from their ropy umbilical cords and knobby limbs to their tufts of hair and tightly closed fists.

The other depiction is the last one I find, and I spend a long time looking at it. It's by Leonardo da Vinci, a simple sketch in soft charcoal from around 1505. It shows the uterus opening like a locket; inside is curled a lovingly rendered fetus, clasping his knees and turning his face away from us. All we see are the curves of his thigh and his shoulder, the back of his smooth, round head. As it turns out, da Vinci got many of the technical details surprisingly correct; as an anatomical illustration, "it was not bettered for more than two centuries," one perinatal expert notes. But more than that, da Vinci got the *feeling* right—as if acknowledging how little men knew, and how closely women clasped their secrets to their bodies.

Scientific knowledge of the fetus advanced haltingly in the centuries after da Vinci's uncannily precise drawing (apparently based on remains from an autopsy). Progress was hindered not least by concerns about propriety; debates broke out among doctors about the morality of examining a pregnant woman without her corset. In the late eighteenth century, physicians like the German doctor Willhelm Gottfried Ploucquet permitted themselves to palpate the bellies of pregnant women—but only those who were pregnant out of wedlock, poor and desperate, who traded such indignities for a bowl of soup or a place to stay for the night. "The movements of the fruit" in the womb, wrote Ploucquet in 1788, "in as much as they can be felt by an outsider or when they can be seen from the outside and are visible to the examiner, are one of the most privileged signs of pregnancy." A few decades later, in 1821, French doctor Jacques Alexandre Le Jumeau de Kergaradec first put a stethoscope to a pregnant woman's abdomen. He was listening for what he imagined would be the fetus "splashing" in amniotic fluid, cavorting like a sparrow in a birdbath. Instead he heard a sound he likened to the movement of a watch: the fetus's heartbeat, twice as fast as its mother's.

Inevitably, the heart rate of the fetus became one more way to guess its sex: a male fetus's heart was said to beat faster, a female's, slower. This turned out to be another wrongheaded theory, though it persisted well into the twentieth century. By then, yet another reason for wanting to know fetal sex had emerged: a growing understanding of the genetic basis of sex-linked diseases such as hemophilia, the bleeding disorder that affects males much more often than females. If pregnant women could learn the sex of their babies before birth, they would have an early clue to whether their offspring would suffer from the dreaded condition.

In 1949, Canadian cytoembryologist Murray Barr discovered that the nuclei of cells from female humans contained distinctive structures called sex chromatins (now commonly referred to as Barr bodies), while cells from males did not. The opportunity opened by Barr's discovery was clear: if you could detect the presence or absence of Barr bodies in cells from the developing fetus (procured through amniocentesis), you could determine the sex of the fetus well before birth. Researchers around the world got

116

to work, and in one of those striking feats of scientific convergence, a number of them arrived at their common goal at the end of 1955. Within the space of a few weeks, four different teams had realized a dream held by humankind for thousands of years.

Nothing so important as a royal succession or a fatal disease hangs on whether I'm having a boy or a girl. I just want to begin knowing this baby, imagining the shape of my future family—to linger over a frilled dress or a striped pullover, to doodle names in the margins of my notebook like a lovestruck teenager. I'm moved by simple curiosity.

Curiosity is also what drove a young doctor named Ian Donald. Donald was passionate and impulsive, sharp-witted and quick-tempered; an American colleague called him a "mad, red-headed Scotsman." A skilled teacher and a sympathetic clinician, Donald also liked to tinker with machines. "I've always been an engineer at heart, I suppose," he once reflected. "My wife says it's because I only had daughters—no sons—and therefore nobody to play trains with."

Donald brought his engineer's sensibility to his work as a professor of midwifery at the University of Glasgow, where he joined the faculty in 1954. His aim was to take a picture of the fetus, and he thought he knew how to do it: sonar. During World War I, the French physicist Paul Langevin had developed the technology of sonar to counter the threat posed by Russian submarines. Sonar works by sending high-frequency sound waves through water; the waves bounce back an echo when they encounter a solid object. To Donald, such an approach had obvious applications to the human body. As he put it: "There is not so much difference, after all, between a fetus in utero and a submarine at sea."

Early investigations of the medical use of sonar had required that the subject be dunked in a giant tank of water, not an appealing prospect for Donald's pregnant patients. In his efforts to dispense with the tank, Donald undertook a series of experiments that quickly became a comedy of errors. First he tried dipping the ultrasound probe in a bucket of water balanced on the patient's swelling abdomen. "As may be imagined, such a situation, in the patient's own bed, had all the ingredients of a shambles, and accidents resulting in wet beds were frequent," Donald reported.

Next he came up with the idea of a water-filled balloon, and determined that condoms were just the right size and shape for the job. A visiting professor was cajoled into buying the condoms while Donald waited in a car outside the shop. When asked by the clerk which kind of condom he wanted, Donald related, the professor "said he did not know but would run out to the car and find out. This story," he added ruefully, "soon received worldwide circulation."

At last, through his indomitable efforts, Donald achieved the view into the womb he had long sought. "I will not forget the excitement with which I observed the first very early pregnancy of about six to seven weeks," Donald recalled. "Now, at last, we could really study pregnancy from one end of its development to the other." For all of human history, he observed, "the developing fetus has been hidden behind a veritable iron curtain." Now, he had drawn back that curtain to reveal what lay inside. The pictures he produced were shown to the world for the first time in June 1958, published in the medical journal the *Lancet*.

I study these images now. The scans show the outline of a fetal skull, a perfect oval; a pair of twins, bodies nestled close; a uterus occupied by a fourteen-week-old fetus. (In a caption accompanying this last picture, Donald noted that before the ultrasound was performed, doctors had misdiagnosed the pregnancy as a fibroid tumor.) The images are indistinct, hard to decipher, like cloud formations against a dark sky. They remind me of the grainy footage beamed back from the first moon landing. By showing us the inside of the uterus, Donald changed the way we look at pregnancy, a shift in perspective as radical as seeing the Earth from outer space.

Some doctors, it's true, were not impressed. One rival in Edinburgh sniped that Donald had used a machine costing ten thousand British pounds to do the job obstetricians routinely accomplished with a two-penny rubber glove. But the pioneer of obstetric ultrasound had no doubts about the significance of his innovation. Indeed, it led Donald, who died in 1987, to a prescient insight. "The first forty weeks of existence," he declared, "may well prove to be far more important medically than the next forty years."

Today, thanks to Donald's invention and its descendants, we take access to the fetus for granted, even in early pregnancy. Now that I'm five months along, a doctor equipped with a stethoscope might be able to detect my fetus's heartbeat—but I already heard it myself more than two months ago. At my twelve-week visit, when no one but my husband and my obstetrician knew I was pregnant, the OB rolled a Doppler device over my midsection. Immediately a sound filled the room: a fast, regular beat beneath an aquatic gurgle, like a horse cantering underwater.

In a sense, technology has even moved the date of quickening. At that same doctor's visit, the obstetrician applied an ultrasound probe to my belly and flicked on a monitor. There was my fetus moving on the screen, limbs jerking loosely like a marionette. I watched, fascinated and delighted, but oddly detached. I felt like a spectator to my own body, like someone watching a tele-vised report of an event unfolding inside her own house. Social historian Barbara Duden has called this experience a "techno-logical quickening": fetal movement detected by ultrasound but imperceptible to the woman herself. It's a neat reversal of author-ity. Once upon a time, doctors relied on women for information about the fetus; today women must consult their doctors to find out what's happening inside them. Actual quickening, Duden writes—once so momentous, even transformative—is now "an experience that has lost its status."

There's no doubt that seeing one's future child on an ultra-sound monitor is a powerful experience, the first visual evidence of the fetus in a culture in which seeing is believing. The encoun-ter can be so compelling, in fact, that some medical providers are using it not just as a diagnostic tool, but as a treatment in itself. Zack Boukydis, a professor of psychology at the Illinois Institute of Technology, has been working with doctors for almost a decade to expand routine screenings into lengthier "ultrasound consulta-tions," in which the pregnant patient is encouraged to point out her fetus's features, observe its behavior, and notice its responses to actions of her own such as talking or laughing. "I was inspired by the pediatrician T. Berry Brazelton, who popularized the idea of 'touchpoints,' or critical moments in child rearing," Boukydis says. "It seemed to me that seeing the fetus on the ultrasound had become an expectant mother's earliest touchpoint, her very first experience as a parent."

He and his colleagues developed a training program for sonog-raphers, and tested the effects of the enhanced ultrasound on pregnant women's psychological state. Those who participated in the expanded session, Boukydis found, displayed reduced anxi-ety and greater attachment to their fetuses; they tended to agree strongly with statements like, "After watching the ultrasound I know my baby better," and "After viewing the ultrasound I feel more attached to my baby." The psychologist is currently studying the use of such consultations with diabetic pregnant women, on the theory that they may become more motivated to stick with their medical regimens, and with depressed pregnant women, who might be induced to feel more positively about themselves and their pregnancies.

Researchers like Boukydis, and often doctors and pregnant women themselves, now commonly speak of the role of ultrasound in helping expectant mothers "bond" with their fetuses. We've become so accustomed to this notion that we may not recognize how peculiar it is, and how new. When the notion of maternal bonding was first introduced, fewer than forty years ago, it referred to the then-novel practice of allowing new mothers to spend time with their babies immediately after birth. Women who touched and held their infants during this sensitive period, some researchers contended, demonstrated better parenting skills, and their offspring scored higher on developmental tests. Now, this process is expected to begin even earlier, before the baby has exited the womb; what the pregnant woman bonds with is not the smell or feel of a naked newborn on her chest, but a spray of pixels on a screen. The clear improvements in visual technology have created an emotional muddle: its images generate intimacy and distance at the same time. A technological quickening fills the mind with images, but it can leave the other senses hungry.

And indeed, it's a distinctly different experience I have one evening early in my fifth month. I'm lying in bed, balancing a magazine on my midsection, when I feel a gentle but insistent prodding—from the inside. The movement stops, and I lie motionless, not breathing, waiting. Then it comes again, like a firmer second knock at a door. In a rush I realize I'm not alone in my body; my baby is here with me. My eyes fill with unexpected tears, and for a moment I feel like Lady Castlemaine, crying out in surprise at a fancy dinner: I am undone.

A few days later a package arrives for me in the mail, courtesy of eBay. It's an old issue of *Life* magazine, yellowed and brittle. The paper crumbles a bit in my hands, but the picture on its cover delivers the same visceral punch as when it first appeared more than forty years ago. The photograph is of a fetus, close up and in full color; its eyes are softly closed, its hands clasped to its chin, its knees gently bent. "Drama of Life Before Birth," trumpets the cover line; inside there are more pictures, and an essay that begins, "This is the first portrait ever made of a living embryo inside its mother's womb."

The pictures were taken by the Swedish photographer Lennart Nilsson, and published in *Life* in April 1965. They caused a sensation: all eight million copies of the issue sold out, the only time in the history of the magazine. Nilsson's images were a world away from Ian Donald's cryptic ultrasound scans, taken just seven years earlier. "Using a specially built super wide-angle lens and a tiny flash beam at the end of a surgical scope, Nilsson was able to shoot this picture of a living fifteen-week-old embryo, its eyes still sealed shut, from only one inch away," marveled the *Life* editors about one of the pictures. The fruit of the womb, so long shrouded in darkness, could now be scrutinized by readers in broad daylight.

But even as Nilsson brought the fetus out of the shadows, he made the pregnant woman disappear. The fetus on the cover appears to be floating in a translucent bubble, its umbilical cord stretching off to the side; surrounding the tiny astronaut's capsule there is darkness, the cool void of outer space. Critics have since pointed out some inconvenient facts about how these images were produced. By selectively enlarging and cropping his photographs, Nilsson made his fetal subject appear solitary and self-sufficient, when in reality it would have been intimately connected to a woman and entirely dependent on her for its survival. Moreover, only one of the pictures in the *Life* magazine spread was of a living fetus; the rest, including the one on the cover, were of fetuses whose bodies were obtained by Nilsson following spontaneous or surgical abortions. "Manipulation was crucial in the construction of the famous cover image," notes British scholar Clare Hanson. "By placing the embryo against the background of a starry sky and isolating it almost completely from the maternal environment, Nilsson created an iconic image of the embryo as spaceman, a heroic figure of pure potential. The physical reality of the maternal body was elided and in its place was a figuration of the womb as empty space."

Manipulated, even deceptive, though it may have been, Nilsson's "iconic image" set the pattern for the fetal portraitists who were to follow. Their efforts fill a stack of books on my desk; leafing through them, I see Nilsson's spaceman everywhere: head big and round as a cosmonaut's helmet, drifting in its own private orbit. The more recent photographs are even more detailed than Nilsson's pioneering work: each tiny fingernail picked out, each vein visible through translucent skin. But for all their scrupulous verisimilitude, these photographs are missing the essential fact about a fetus, the precondition of its existence: its connection to a pregnant woman. Without her in the picture, these images are no more than an optical illusion.

Such still pictures were soon followed by videos made in the womb. One evening I slide a series of discs into my DVD player, in what is surely one of the stranger movie marathons ever staged. I watch as film after film enacts the same drama, variations on a tried-and-true formula: sperm meets egg. Following fertilization, the embryo moves via time-lapse photography through the form of a bug-eyed fish, then a spindly bird (shades of Haeckel's theory of recapitulation); finally it plumps and swells into something like a baby. The narrator of one movie, I note, here makes the familiar claim for ultrasound: "Doctors have discovered that for parents-to-be, the scans play a beneficial role in developing emotional bonds," he intones. "Research has shown that seeing the face and expression of the developing fetus while it's still inside the womb can be an intense bonding experience that can provide a boost to the baby's development once it's born, and also for the long-term relationship between the child and her parents."

In these films the camera revels in its virtuosity, swooping through Fallopian tubes, piloting the perimeter of the uterus. The images are awe inspiring, and often beautiful. But after a while I begin to feel uncomfortable; I fight the urge to look away. There's something intrusive in the way the camera elbows into the womb, inspecting every inch of the fetus's body. I wince as it zooms in close on an unsuspecting face; the fetus seems to start, like someone waking from a deep sleep to the clicking of camera shutters and the popping of flashbulbs. As the camera pans the length of the fetus's body, I feel like a peeping Tom, seeing things I shouldn't see, gaining knowledge I shouldn't have. While I watch, the fetus's position shifts slightly, and its genitalia come into view: the grape leaf lifted, her final secret revealed.

*   *   *

Today, the old ways of guessing male or female have mostly moved aside for more official determinations: as many as 80 percent of women choose to find out the sex of their babies through amniocentesis (usually in the fourth month of pregnancy) or ultrasound (around the fifth month). I well remember getting the results from a sonogram in the middle of my first pregnancy. Afterwards John and I lingered for a while outside the clinic, squeezing each others' arms, shaking our heads, laughing the giddy laugh of released suspense. "A boy!" we kept saying. "I can't believe it!" Finally we parted, he to his office uptown and I to our apartment in the Village. Alone in the back of a cab, I turned over this new piece of information, this scientific verdict that made all the folklore obsolete.

Or maybe not. Recently, a number of studies have turned up indications that there's something to those old superstitions. Women who are pregnant with girls may indeed experience more nausea, for example: several large investigations have found that women afflicted with severe morning sickness, called hyperemesis gravidarum, in the first trimester are more likely to be carrying a female fetus. Doctors, concluded one study, "can say with confidence to a woman who has hyperemesis gravidarum that she has a 55.7 percent chance of delivering a girl." The sicker she is, the more likely this is to be true: a 2004 study by epidemiologists at the University of Washington found that women who were extremely ill—hospitalized for three days or more—had odds of having a girl that were 80 percent higher than those of women who did not experience severe nausea. A hormone called human chorionic gonadotropin may be to blame: female fetuses produce more of it than males.

Likewise, pregnant women with bigger appetites may well be carrying male fetuses. Obstetricians have long known that boy babies tend to be bigger than girl babies at birth, by an average of 3.5 ounces. But they weren't sure how this difference came about. A 2003 study tracked the diets of more than three hundred women receiving prenatal care at Beth Israel Hospital in Boston. Women who were pregnant with boys consumed about 10 percent more calories than those who were pregnant with girls—more protein, more carbohydrates, and more fats, adding up to about two hundred extra calories per day. Testosterone secreted by fetal testicles, the researchers speculate, could be sending mothers a signal to eat more. (Another possibility occurs to me: perhaps women who learn that they're carrying male fetuses are already obeying the cultural dictate to feed growing boys.)

Finally, one small but intriguing study found that, "contrary to expectations," women who rely on dreams and emotions to guess their babies' sex have a surprisingly good chance of being correct. Among a group of forty-five well-educated pregnant women—all of who had chosen not to learn fetal sex from pre-natal tests—seventeen said they had a "feeling" about whether their fetus was male or female, and thirteen were right. Eight of the women reported having a dream about their fetus's sex, and every single one of them was on the money. "It is always possible that this was a spurious finding," acknowledged the authors, from Johns Hopkins University in Maryland. "It is equally likely that there is simply much about the maternal-fetal connection that we do not know."

In any case, it seems that the question of fetal sex is too important to leave to the experts. The taxi pulled up to my building, and I paid the driver his fare; he turned to look at me as I heaved myself out of the backseat.

"By the way," he said over his shoulder, "you know you're having a boy?"

*   *   *

We may learn the fact of fetal sex earlier than we used to, but it's still long after the decision was made. The sex of the baby I'm carrying was determined five months ago, at the moment of conception; whether it is female or male depends on if the sperm that penetrated the egg was carrying an X or a Y chromosome. Fetal sex would seem, then, to be the ultimate inside job: an entirely internal event, well insulated from outside forces. But consider the fact that only an estimated 20 to 40 percent of fertilized eggs actually results in a live birth; the rest are lost to miscarriage, often before the woman even knows she's pregnant. That leaves plenty of room for other factors to influence whether an embryo, male or female, survives to be born. And indeed, a quirky body of research has grown up around the study of shifts in the sex ratio and their causes. John Arbuthnot's purported "exact balance" between male and female births turns out to be more variable than he knew.

For example: On December 5, 1952, a "killer fog" descended on much of London. The fog—a thick blanket of coal smoke held in place by unusual weather conditions—was so dense that Londoners couldn't see their own feet, or more than a foot in front of their noses. The fog seeped into schools and opera houses and the stacks of the British Museum; it even penetrated people's clothes, blackening their undergarments. It lasted for four days, and killed at least four thousand people. But the oddest consequence of the "Big Smoke" of 1952 arrived some months later, when an unexpected development was noted: fewer boys than expected were being born. At sixteen large London hospitals, a total of 144 female births were recorded for the third week of October 1953, compared to 109 male births.

A random anomaly, one might think—but researchers have documented the same phenomenon following economic depressions, natural disasters, and political and social upheavals. In the wake of these stressful events, the sex ratio—that is, the ratio of male births to female births, which normally hovers around 105 boys to 100 girls—drops. It happened after the East German economy collapsed in 1991: While fewer boys were born in the suddenly poverty-stricken East Germany, the sex ratio remained normal in still-thriving West Germany. It happened after a devastating earthquake struck Kobe, Japan, in 1995: About nine months after the quake, the sex ratio fell in the hard-hit Hyogo Prefecture. It may even have happened after terrorists flew planes into the World Trade Center towers on September 11, 2001: Fewer boys than normal were born to women in New York City in the months after the attacks.

A stress-related decline in the sex ratio is no anomaly, but a biological mechanism that evolved to keep the species going, according to Ralph Catalano, professor of public health at the University of California, Berkeley. "From an evolutionary perspective, in stressful eras it's a smarter bet to have a female child than a male child. A daughter is more apt to provide you with grandchildren," Catalano tells me. In hard times, he explains, weak males are less likely to survive and to reproduce, while females are more likely to carry on the family line. Women who are carrying a weak male fetus would do better, from a Darwinian point of view, to end that pregnancy and start another, this time carrying a female or a stronger male.

How this might happen is a matter of speculation. It may be that in times of great stress, hormonal changes in the bodies of pregnant women cause the spontaneous abortion of weak fetuses; male fetuses are especially likely to be lost, because they tend to be less robust to begin with. Catalano found support for this theory in a study he made of the life spans of Swedish men born between 1751 and 1912. Men born in years when the sex ratio was low tended to live longer than men born when the sex ratio was high. In Catalano's interpretation, published in the *Proceedings of the National Academies of Science* in 2006, this fact suggests that in stressful times weak male fetuses are "culled" by miscarriage, leaving fewer, but stronger, males to be born.

As Catalano imagines it, the womb is not a tender cradle of care, but an arena of ruthless selection and struggle for survival. "Now that almost all children in the developed world live to adulthood, most of the natural selection that goes on today happens in utero," he says. In this unforgiving sphere, males are already at a disadvantage. "Compared to men, women are biological fortresses," says Catalano. "There's no doubt that men are the weaker sex, starting as early as the womb."

No one is more keenly aware of the vulnerability of males than the inhabitants of Sarnia, a small town in Ontario, Canada, that is home to a community of Native Americans called the Aamjiwnaang. A few years ago, members of the tribe began voicing concern that fewer and fewer boys were being born. Women in expectant-mother groups noticed that most of their members were having girls; local coaches could not find enough boys to field a baseball team. A study led by Constanze Mackenzie, a professor at the University of Ottowa Medical School, confirmed that male births among the Aamjiwnaang had begun falling in the early 1990s, and that the decline was accelerating. By the early 2000s, reported Mackenzie, nearly twice as many girls as boys were being born to the women of Sarnia.

A community health survey investigating potential causes of the phenomenon is now underway. But the decrease in male births occurring in Sarnia may also be happening on a less dramatic scale in other parts of the world. Sex ratios in a number of industrialized countries appear to be dropping. In the United States, for example, there were 1,059 male births for every 1,000 female births in 1946, according to the Centers for Disease Control. In 2001, there were 1,046 male births per 1,000 females, the lowest sex ratio in six decades. Similar drops have been reported in the United Kingdom, Canada, Japan, Finland, Norway, Denmark, and the Netherlands. These changes in the sex ratio could represent a "sentinel health indicator": an unusual pattern in the health of a population that serves as a warning signal, a kind of canary in the coal mine alerting us that there's something amiss.

I first hear this phrase from Devra Davis, an epidemiologist and environmental health expert at the University of Pittsburgh. The change in sex ratios is small, but real and consistent, she tells me. "We're talking about the reproduction of the species here," says Davis. "Anything that disrupts that is a matter of concern." In 2007, Davis carried out a study that found that the sex ratio declined significantly among white Americans (though not among African Americans) between 1970 and 2002. Over that thirty-two-year period, the decline was the equivalent of a shift from male to female births of 135,000 white males. Why is this happening? "The short answer is, we don't know—and it's almost certainly not just one thing," Davis says. "But a leading explanation is that exposure to environmental chemicals is affecting prenatal development, perhaps making male embryos and fetuses more vulnerable to miscarriage." The town of Sarnia, she notes, is nearly surrounded by large petrochemical, polymer, and chemical industrial plants.

A number of other explanations for the decline in male births have been proposed, some more plausible than others. It could be the everyday stress we're under (as opposed to the stress of an event like a natural disaster). Research has found that female animals who are stressed have fewer male offspring, and human studies have suggested that the same may be true for us. In a study of more than six thousand Danish women evaluated for psychological distress in early pregnancy, for example, 47 percent of those who were most distressed gave birth to boys; among those who were least distressed, 52 percent delivered boys. An ingenious study by Ralph Catalano, generated by an examination of medical records in Sweden, found that the more antidepressants are prescribed to women (a gauge of the stress they're experiencing), the fewer boys are born.

It could be single mothers: A 2004 study of more than eighty thousand births over forty years found that boys are slightly less likely to be born to women who are not living with a man at the time of conception. The study's author, epidemiologist Karen Norberg of Washington University, notes that male offspring require a greater investment of calories even as fetuses (never mind as teenagers), and so gestating and raising a male child could be a greater risk for a woman with only her own resources to draw on.

Or it could be skipping breakfast. In 2008, Fiona Mathews of the University of Exeter and two colleagues from Oxford University announced in the *Proceedings of the Royal Society* that they'd found a relationship between women's nutrition around the time of conception and the sex of their babies. In particular, women who ate cereal every morning had a higher chance of delivering boys. Perhaps filling meals like these send a signal to women's bodies that food is abundant, Mathews speculated, and so having a male child is an acceptable risk. The fact that many young women of reproductive age skip breakfast or consume low-calorie diets in order to keep their weight down, she added, could help explain the falling rate of male births.

Mathews's study soon came in for a scathing rebuke from Stanley Young, a statistician at the National Institute for Statistical Sciences, a U.S. government agency. "Cereal-induced gender selection?" he scoffed in a published reply to Mathews's article. Rather, Young argued, Mathews's finding "is easily explained as chance." If you ask study subjects enough questions, he points out—and Mathews asked pregnant women about their consumption of more than a hundred different food items—you will find some statistically significant relationships simply by chance. Results like these, Young concludes, "should be viewed with some skepticism, since the human imagination seems capable of developing a rationale for most findings, however unanticipated." For her part, Mathews defended her hypothesis as "based on sound evolutionary principles and supported by considerable research on other mammal species."

Perhaps these proposed explanations have merit, or perhaps they're modern-day versions of the folklore that surrounds so much of pregnancy, and fetal sex in particular. For now, I file them away, along with the fact that I eat a bowl of raisin bran every morning.

Such influences on fetal sex, if indeed they exist, are accidental and unwitting, but humankind has long wished to deliberately choose the sex of children. In ancient Greece, this meant lying on one's right side during copulation; in eighteenth-century France, it meant binding up the man's left testicle before intercourse; in twentieth-century Italy, it meant biting the woman's right ear during sex (following the familiar bias, all these methods were meant to ensure the birth of a boy). In today's America, it most likely means reading Dr. Shettles.

Back in 1956, Landrum B. Shettles was among the researchers who announced that they could determine the sex of a fetus by analyzing amniotic fluid. But Shettles, a professor of obstetrics and gynecology at Columbia University, wanted to take things a step further: Would it be possible not just to identify fetal sex, but to select it? As a book by Shettles and a coauthor, David Rorvik, told the story, couples often approached the doctor to confide their desire for a girl or a boy: "They imagined he might be sympathetic to their longings—and they were right! Of course, they couldn't know at the time that Dr. Shettles had been asking himself the same question, a question that he would eventually answer not only to the satisfaction of his patients and hundreds of thousands of others around the world, but also to his own satisfaction as he went about creating a family of his own, consisting of three boys and three girls!"

The book is *How to Choose the Sex of Your Baby*, and it has sold more than 1.5 million copies since its publication in 1970. In it, Shettles presented his idiosyncratic take on the facts of life: sperm carrying the male chromosome swim faster and die sooner, he contended, while sperm carrying the female chromosome move more slowly and survive longer. Therefore, the sex of the child conceived can be controlled by manipulating the timing of intercourse and the relative acidity or alkalinity of the vagina. Women wishing to have a boy, Shettles counseled, should wait until the day of ovulation to have sex, douching with baking soda beforehand; those wishing to have a girl should have sex three days before ovulation, douching with white vinegar. Shettles trumpeted a success rate as high as 85 percent among women who faithfully followed his instructions. Other scientists were dubious, to say the least. The chair of obstetrics and gynecology at Columbia, worried that Shettles was making a "laughingstock" of the department, even issued a press release distancing himself from Shettles's claims.

Though scientists have generally not deigned to test popular theories of sex selection, an exception was published in the *New England Journal of Medicine* in December 1995. Allen Wilcox, a scientist at the National Institute of Environmental Health Sciences, recruited more than two hundred women who were trying to become pregnant, asking them to keep daily records of when they had sex and then following them until they gave birth. His conclusion: "For practical purposes, the timing of sexual intercourse in relation to ovulation has no influence on the sex of the baby." When I get Wilcox on the phone, he is even more blunt: "There's zero evidence that the Shettles method works," he told me. "And that goes for all the other folk wisdom out there, too." A survey conducted by researchers at the University of Alabama bears Wilcox out: it found that 8 percent of the 243 pregnant women surveyed had tried to influence the sex of their baby, using methods like sexual position, timing of intercourse, and "following guidelines in a magazine article on how to conceive a male fetus." Their success rate: 47.7 percent, worse than if they had relied on chance.

The Shettles method was once put to a real-life test, at a clinic in Singapore in the 1970s. "The clinic was set up in the hope that if couples could have the son they desired, they would keep their families small in accordance with the government's strong commitment to population control," wrote Nancy Williamson, an American scientist who observed the experiment. All thirty-one women who enrolled wanted boys. "By chance alone, one would expect about sixteen to have boys," Williamson noted. "In fact, fourteen had boys." In other words, the Shettles method had failed miserably. Williamson concluded her report with a simple suggestion: "A better solution would be to discourage boy preference and improve the opportunities for girls."

Williamson's wise advice went unheeded. In the years after her study was published in 1978, ultrasound became increasingly available in Singapore and other countries in South and East Asia. The introduction of this technology enabled a new and brutal form of sex selection: if the sonogram shows it's a girl, abort. Researchers' estimates of the number of female fetuses aborted over the past twenty years are mind-boggling: ten million in India, twenty million in China. Because of this practice, demographers and political scientists point out, millions of males in these countries are now reaching adulthood with no prospect of finding a mate and starting a family. In the absence of such social bonds, they warn, these rootless young men may turn to violence, militarism, or even terrorism. Here we have perhaps the ultimate example of how events within the womb can have wide-ranging repercussions in the world: an intervention in *there*, repeated millions of times, is reshaping societies out *here*.

Recognizing that the lopsided sex ratio could spell social disaster, some Asian governments have outlawed sex-selective abortion, and have even made it illegal for a pregnant woman to find out if she's carrying a boy or a girl. Still, a strong and persistent preference for boys has made the practice exceedingly difficult to stamp out—but before passing judgment on these societies, consider the evidence that sex-selective abortion goes on in the United States as well. Researchers at Columbia University found in a 2008 analysis of Census Bureau figures that among American-born children of Chinese, Korean, and Indian parents, the sex ratio for first children is normal. If that first child is a girl, however, the odds of the next child being a boy go way up, and rise higher still for a third child who already has two sisters. "We interpret the found deviation in favor of sons to be evidence of sex selection, most likely at the prenatal stage," the authors conclude.

In just the past few years, technology has contributed two more morally palatable ways to choose fetal sex. The first: create embryos in a Petri dish, and implant only those of the desired gender. Forty percent of U.S. fertility clinics already offer this option to their clients, and almost 10 percent of preimplantation genetic diagnoses, as the analyses of the embryos are called, are performed for gender selection alone. The second technique involves sorting sperm according to whether they carry an X or Y chromosome (using a machine originally developed to sort the sperm of cattle); the sorted sperm are then used to artificially inseminate the woman, or to create embryos that are then implanted. The technology is available only on an experimental basis, but so far more than a thousand babies have been born this way.

As the ability to choose a boy or a girl becomes still easier and less ethically fraught, will more of us seize this power for ourselves? In a survey of more than one thousand Americans, respondents were asked if they would choose the sex of a child if it were possible to do so simply by taking a pink pill for a girl and a blue pill for a boy; 18 percent said yes, they would. Of course, sex selection isn't nearly so simple, but give it time. With time, attitudes may change as well, becoming as accepting of sex selection as we now are of formerly startling notions like in vitro fertilization and prenatal testing for genetic defects.

Even if we choose not to intervene, we'll be finding out about fetal sex earlier and earlier. Sex identification kits now sit next to pregnancy tests on drugstore shelves; the marketing materials for one such product, the Intelligender Gender Prediction Test, boast that it "bridges the curiosity gap between conception and sonogram." That ever-shrinking curiosity gap means that we will be applying our expectations and assumptions about gender to our children sooner than ever before. Sociologist Barbara Katz Rothman observed the beginning of this phenomenon more than two decades ago in her 1986 book about the impact of amniocentesis, *The Tentative Pregnancy*. Rothman asked the 120 pregnant women she interviewed to describe the movements of their fetuses. Women who'd learned they were having a girl gave answers such as "very gentle, slow, more rolling it seemed than kicking," "moderate, reassuring but not violent," "quiet in the mornings and afternoons," "lively but not excessively energetic." Mothers who knew they were carrying a boy described "many somersaults and very vigorous movements," "rolling from side to side and little kicks and punches up and down," "a constant jabbing under my ribcage," and "a saga of earthquakes." Tellingly, the responses of women who did *not* find out the sex of their fetus showed no such stereotypical patterns. "I used to say, along with most social scientists, that gender socialization begins at birth," Rothman wrote. Now it begins, like so much else, long before.

Two hundred years ago, no one had yet heard a fetal heartbeat through a stethoscope; fifty years ago, no one had yet seen a fetus on an ultrasound screen. As technology advances, less and less about the fetus is left to the imagination. Sometimes, this new knowledge feels distancing, or even exploitative; sometimes it allows us to impose our biases, or even to act on them. What it does not do is extinguish our imaginings. Indeed, it can seem as if each new window into the womb provides another opening for our wishes and fears to present themselves, a temptation to which even scientists are not immune.

Take, for instance, the use of three-dimensional sonography to study the fetus. Patented in 1987, high-speed 3-D ultrasound offers the clearest picture yet of the fetus in the womb; the sonographer can measure not only the size but the volume of fetal organs and other structures, and can zoom in on a "slice" or cross-section of the fetal body. Now making its way into everyday obstetrical practice, it is also being used by researchers—to investigate fetal abnormalities, for example, but also to study nuances like fetal facial movements, even "expressions." A team of Japanese scientists, led by Toshiyuki Hata of the Kagawa University School of Medicine, is using 3-D ultrasound to count fetal blinks and yawns, and to document what they call expressions like smiling, scowling, and sticking out the tongue. A 2006 paper published by Hata and his colleagues in the *International Journal of Gynecology and Obstetrics* offers a series of images to illustrate their work, a gallery of fetal faces. The expressions they describe are easy to identify: in one picture, a fetus's mouth stretches wide, turning up at the corners; in another, a fetus's brows are drawn together, and the corners of its mouth are turned distinctly down.

But the attribution of emotion that accompanies words like "smiling" and "scowling" seems dubious when applied to a fetus; as weary new parents know, even babies don't smile until they're at least a month old. Hata and his colleagues dutifully note that the study of fetal facial expressions could one day be used to evaluate development in the womb: 3-D technology, they write, "may be a key to predicting fetal brain function and well-being and an important modality in future fetal neurophysiologic research." Their real interest in fetal faces seems to lie elsewhere, however. Beneath their scientific justifications, one can almost hear the excitable exclamations of a parent: "Look! Is she smiling? I really think she's smiling!"

On a November morning in the middle of my fifth month, I'm squeezing John's hand hard enough to stop his circulation as we both stare, riveted, at a monitor mounted high on the wall. We're here at the sonography clinic to get a fetal anatomical survey, a comprehensive ultrasound examination that will reveal whether my fetus is developing properly—and, incidentally, whether it's a boy or a girl. This piece of information seems anything but incidental to me, however: I feel my impatience rising as the technician slides and clicks her mouse, clicks and slides, measuring the length of a femur, the radius of a skull.

"Here's the heart . . . here's the liver . . . ," she murmurs. John and I crane our necks, trying to identify said organs amid the shifting shadows on the screen. "Wait'll you see this," the sonographer says suddenly. She pushes a button, and the colors on the monitor change from ghostly grays to shades of burnt umber. "Your fetus in 3-D," she announces. All we see at first are the folds of the uterine wall, rippling and billowing like a curtain. The probe on my belly pokes, pauses, jiggles—and out of the folds pops a hand: perfectly articulated, three-dimensional, waving vigorously. John and I rear back like the audience at a 3-D showing of *Jaws*. The hand is so lifelike I feel that I could take it in my own.

With a few more nudges of the ultrasound probe, my fetus's whole body comes into view: it looks sculptural, molded, as if shaped by human thumbs. Rendered in the color of clay, it reminds me of the models made by Italian artisans I'd gazed at in the library. John has a less elevated association. "It looks like a fetus in Claymation," he whispers, calling to mind the pliable green Gumby. The technician turns from the screen to face us. "Do you want to know the sex of your baby?" she asks. I nod, and close my eyes. John squeezes my hand back.

It's no doubt true that less is left to the imagination these days; while the women of previous generations had pictures only in their minds' eye, we have fetal snapshots posted on our refrigerators. But many of the mysteries that have made pregnancy the object of wonder and speculation for millennia are still intact. We will always want to know more about our fetuses; our knowledge will always be limited; and our imaginations will always rush in to fill the gap.

The technician is quiet for a moment; all I can hear is the click-click-click of her mouse. Behind my eyelids, I see an image of John and his younger brother, tossing a ball in their backyard; I see my sister and me, braiding each other's hair in thick ropes down our backs. In this moment before the sonographer speaks, my baby could still be anything: a girl, a boy, a girl . . .

"It's a boy," she says.

SIX MONTHS

My name for it—admittedly a bit melodramatic—was Mommy Despair. It arrived in the exhilarating, exhausting weeks after my first son's birth, along with its louder male counterpart, Daddy Rage. John, usually so even-tempered, would be provoked to helpless fury by impossibly complicated baby carriers and cribs with too many parts; these episodes of epic frustration would invariably end with him lying, defeated, on the floor beside the triumphant piece of equipment. Daddy Rage eventually subsided as John mastered the ways of Allen wrenches and BabyBjörn straps, and as he righted himself in the upside-down world of caring for a small child: drowsy in the daylight, wide awake in the dark; days that speed past without a single task or conversation completed, full of hours slow enough to hear each second's tick.

My affliction was quieter, darker, more lingering. I remember the first time I was left alone with the baby, after the noisy rush of visitors receded like a tide. The apartment was very still: just me and Teddy, in his car seat set on the floor at my feet. The dog padded over and sat down next to the baby, and two pairs of wide round eyes gazed up at me, as if to ask, "What now?" I didn't have an answer. My life had changed in ways I couldn't control and didn't yet understand, and my muddled sense of loss was all the more confusing next to the fresh joy I felt in Teddy.

Such emotions were overpowering, but it was the everyday skirmishes that brought me close to surrender, what writer Joan Didion once described as fighting "a guerrilla war with my own life." (Her daughter was two years old at the time.) Over the next few years, I found myself waylaid by absent babysitters, ambushed by colds carried in from preschool, trapped by bad weather in close quarters with an explosively energetic toddler—pitched battles in which I was always outgunned. My writing fell by the wayside, a casualty of the unrelenting demands of motherhood. I became someone I barely recognized: a person who carried packs of wipes and bags of Cheerios instead of pens and a notebook, who dressed to disguise spit-up stains and breast-milk leaks, and who referred to herself in the third person: "Mommy's right here." "Mommy doesn't like that." "Give that to Mommy, now." Confusion turned to anger, and because there was no good place to put it, anger turned to despair.

But then, slowly, I start to return to something like my former self. Learning to write and be a mother is like having to learn to walk again: tentatively, unsteadily, putting one word in front of another. Three years on, I've almost regained my balance—and now the growing weight of my belly threatens to throw me off again. As fall shades into winter I feel Mommy Despair descending again, like a scrim lowered between me and the rest of the world, dulling even the pleasure I take in my pregnancy. At night I have dreams so obvious they require no interpretation: I lose the baby in a tangle of sheets on a giant bed; now I lose my manuscript, its pages gusted out of my hands and falling, fluttering, into a chasm.

When I try, awkwardly, to talk about these feelings, my confidantes—even other mothers—seem taken aback that a pregnant woman would feel anything but simple happiness. Or they smile and touch my arm and say, "You're just hormonal." I have two choices, it seems: to be serenely untroubled, or weepily irrational. What I actually feel is more complicated than either of these caricatures—even more so because of what I don't say: that I'm worried that the gloom coursing through me is making its way to my fetus, that what I'm feeling, he feels too.

Conflicting ideas about women's moods during pregnancy and early childrearing—that it's a time of beatific bliss, and that it's a time of hormonally charged volatility—have vied with each other for centuries. On one side are arrayed the many portraits of the Madonna, radiantly cradling a child or aglow with a child to come. Now a new addition to their ranks has been put forward: researchers have proposed that the Mona Lisa, painted by Leonardo da Vinci in the sixteenth century, was pregnant. Records unearthed in Florence show that Lisa del Giaconda, a leading candidate among the women thought to be da Vinci's subject, gave birth four months after the painting was begun. In his recent biography of da Vinci, writer and Yale University physician Sherwin Nuland embraces the idea, observing that the portrait sitter's face is full and round, that her fingers seem swollen (and bear no rings, unusual for a woman of her station), and that her hands are folded protectively above her abdomen in a manner characteristic of pregnant women. The Mona Lisa, writes Nuland, exudes an "inner satisfaction that the miracle of life is being created within her body"; her iconic expression, he says, "is a smile of the coming of life."

And yet there is also a persistent belief that pregnancy produces emotional instability, even madness. In the nineteenth century, the diagnosis known as "insanity of pregnancy" was freely dispensed. Gestation, the theory went, caused the uterus to become irritable or excitable; other organs, including the brain, followed suit. Such derangement may begin any time after conception, and may end after delivery or continue after birth, wrote physician George Man Burrows in 1828; some women "are insane on every pregnancy or lying-in, others only occasionally." Another doctor, William Thompson Lusk, further described the syndrome in *The Science and Art of Midwifery*, published in 1892: "The marked perturbation of the nervous system" during pregnancy may "give rise to moral perverseness, to the loss of memory, to hysteria, or to hypochondria," Lusk reported. So common are "the more pronounced forms of mental derangement" among pregnant women, he wrote, "that of the insane who crowd the public asylums, in one eighth the malady is of puerperal origin."

Since the time of Hippocrates, these opposing emotional poles have often been the only ones permitted to women in pregnancy. Along with this insistent notion came another: that a pregnant woman's mood, blissful or agitated, can be passed on to the fetus. For many centuries and in many cultures, this process was believed to be quite literal: a shocking sight or strong passion experienced by the pregnant woman would impress itself on her offspring in the form of a birthmark or other physical anomaly. Women who gazed at portraits on the wall during pregnancy might have a hairy or dark-skinned child as a result; women whose minds wandered during intercourse might bear infants who looked more like their lovers than their husbands. Women frightened while pregnant by the sight of an animal would deliver a baby resembling the creature that spooked them; women with strong cravings during pregnancy might have a child with a birthmark in the shape of a fruit or other food.

Popular belief in "maternal impressions" reached its height in the eighteenth century. The phenomenon made frequent appearances in the era's literature: a widely read poem about "how to have handsome children," published in 1709, began: "Ye Pregnant Wives, whose Wish it is, and Care,/To bring your Issue, and to breed it Fair,/On what you look, on what you think, beware." In the 1742 novel *Joseph Andrews*, written by Henry Fielding, a character's strawberry-shaped birthmark is said to be the result of his mother's longing for strawberries while pregnant. In Fielding's *Jonathan Wild*, which appeared three years later, the title character is fated to be a criminal by his mother's "violent desires to acquire all sorts of property" while pregnant with him.

Women's letters and diaries from this time also turn up mentions of maternal impressions, some of which sound strikingly familiar. I feel a start of recognition when I read a letter from Mary Wollstonecraft, the early feminist and author of *A Vindication of the Rights of Women*. It dates from 1784, when Wollstonecraft was five months pregnant with her daughter, Fanny: "It is time for me to grow more reasonable, a few more of these caprices of sensibility would destroy me," she wrote, distraught, to her lover Gilbert Imlay. "I have, in fact, been very much indisposed for a few days past, and the notion that I was tormenting, or perhaps killing, a poor little animal, about whom I am grown quite anxious and tender, now I feel it alive, made me worse."

Other accounts of maternal impressions are so fantastic they pass into the surreal. There's one in particular that I can't get out of my head, a story that could have been penned by the Brothers Grimm, or by Oliver Sacks: the tale of Mary Toft, who gave birth to a rabbit. Toft, a servant woman from the village of Godalming in southern England, had been working in the fields while pregnant when she spotted a rabbit she wished to eat. She became preoccupied with the animal, she said, and her thoughts affected the creature taking shape in her womb. When she went into labor in October 1726, she delivered not a baby but rabbits, sixteen in all. The man-midwife who attended the birth was so amazed by the spectacle that he wrote to several of London's most prominent physicians, who traveled to Toft's village to see for themselves, and came away convinced. Newspapers trumpeted the story on their front pages; Toft traveled to London for a star turn; the Queen's own *accoucheur* was persuaded that Toft's claims were true.

The Rabbit Woman of Godalming was eventually revealed as a hoax, of course. The fraud was exposed when a porter admitted smuggling rabbit carcasses to Toft which she then "birthed," apparently hoping to make a fortune by exhibiting herself and her monstrous offspring. But reading her story now, what strikes me is the willingness of so many to accept it was possible. However preposterous it seems to us today, the acceptance of maternal impressions followed from a belief in the intimacy of the relationship between a pregnant woman and her fetus, and in the centrality of her mental and emotional state. What she contemplated, what she craved, what frightened or excited her—these things were believed to be shared with her fetus, and so were taken seriously by doctors and family members. Pregnant women were shielded from disturbing or frightening sights, and encouraged to expose themselves to pleasant or uplifting ones. (My favorite example: when the pregnant wife of the sixteenth-century German botanist Joachim Camerarius told him that she had an overwhelming urge to smash a dozen eggs in his face, he let her.)

At one time, physicians themselves promoted the doctrine

of maternal impressions. But following the revelation of Mary Toft's deception, a vigorous debate broke out within medicine about whether such accounts were ever accurate. In 1727, doctor James Blondel took a strong stand against the idea, declaring it "a vulgar error" propagated by "tale-mongers" and "imaginationists" on "ignorant people." Blondel, a member of the Royal College of Surgeons in London, made his argument in a pamphlet titled "The Strength of Imagination in Pregnant Women Examined." His denunciation called out a defense from Daniel Turner, another prominent physician with an equally strong belief in the reality of such impressions. It is "made apparent by a multitude of examples," Turner wrote, "how manifest and great an empire the fantasy of a pregnant woman has over the blood and humors together with the spirits of her body, and how by their ministry she is able to give not only monstrous shapes and figures to that of the more tender fetus, but to communicate diseases also."

The two were soon embroiled in a "pamphlet war," trading jabs back and forth. Blondel pointed out that midwives and mothers invoked the notion of maternal impressions only *after* some irregularity was noticed in the newborn child. And, he said, it seemed unlikely that every passing emotion felt by a woman during pregnancy would leave a mark on her offspring. Woman's imagination could not be "so malignant to the fetus as 'tis commonly reported," he asserted, "or else the race of man should insensibly degenerate into a generation of monsters." It was Blondel's view that won the day among the succeeding generations of doctors, eager to achieve greater respectability for their profession by purging it of superstition and folklore. So definitive was their rejection of maternal impressions, in fact, that the notion that a woman's mood might affect the fetus was put aside for more than three hundred years.

*   *   *

In the twentieth century, this age-old idea was taken up again by a discipline on the edge of medicine: psychoanalysis. To the subject of pregnancy, psychoanalysis brought all its justly criticized habits: overconfident assumptions, reflexive misogyny, and extravagant speculation based on little evidence. For example: its claim that pregnant women who experienced morning sickness were expressing a secret wish to be rid of the fetus. Their nausea had not a physical but a psychological cause, some analysts asserted: it was "an oral mechanism of rejection," "a subconscious rebellion against pregnancy." Such women were variously described as "immature, "neurotic," "hysterical," and "frigid," as well as overly attached to their mothers. While the doctrine of maternal impressions had located the danger to the fetus on the outside—some shock or sight that the woman could only do her best to avoid—psychoanalysis offered an explanation more cruelly perverse. Bad things, like morning sickness and even miscarriage, happened during pregnancy because women on some level *wanted* them to. The solution for pregnant women's destructive impulses was, of course, analysis five days a week.

Especially in its early days, psychoanalysis was dominated by men, and by an often-obtuse approach to women's emotions and behavior. Freud himself attributed the desire to be pregnant to a wish to have a penis. (It doesn't seem to have occurred to him that women who become pregnant might actually want . . . a baby.) But there were a handful of women analysts, and they did not exempt themselves from their profession's harsh logic. Helene Deutsch, for example, was an Austrian-born psychoanalyst and writer who was the first of Freud's female students to be analyzed by him. In her massive two-volume work, *The Psychology of*

Women, published in the mid-1940s, Deutsch wrote extensively about the emotional life of pregnancy.

The act of having a child is inevitably accompanied by ambivalence, Deutsch maintained, and the way that a woman resolves these mixed feelings depends on her relationship with her own mother. Indeed, the fate of the pregnancy rests on this "identification"; a woman's emotional state alone could make a pregnancy succeed or fail. To demonstrate this process, Deutsch described in intimate detail the case of a patient she called "Mrs. Smith." Mrs. Smith had repeated miscarriages, which Deutsch attributed to her highly conflicted relationship with her mother. Mrs. Smith was able to give birth to a son, Martin, in 1917, by relying on the emotional support of a close friend, also pregnant, and the friend's warm, loving mother. But after the friend moved away, Mrs. Smith's conflicts reemerged, and she miscarried again.

The pseudonymous Mrs. Smith, Deutsch confessed years later, was really the analyst herself. Knowing this, Deutsch's final judgment on the case strikes me as deeply sad, mixing caustic self-evaluation with genuine bewilderment. "She ironically called herself an 'appendix mother' who could bring her pregnancy to a successful conclusion only by leaning on another woman. Beyond this she was not neurotic and could solve all the other problems of her life," Deutsch wrote. "It was only to the heavy task of pregnancy that she was unequal, for reasons of which she herself became aware. After her friend had failed her she could no longer chase away the shadow of the mother she had rejected." Deutsch added, forlornly, that "psychoanalytic treatment did not remove her difficulties." She never had another child.

Even as psychoanalysis itself has faded as a cultural influence and as a mode of treatment, its punishing philosophy has left a residue of blame and guilt—apparent, for example, in uneasy jokes about "hysterical" pregnant women, and in the common implication that women who miscarry must have done something to cause the loss. But the best of the psychoanalytic literature, including some of Deutsch's, also did women the favor of taking their mental and psychological states seriously. I feel a kind of gratefulness when I come across the work of Grete Lehner Bibring, an Austrian-born analyst who became a professor at Harvard Medical School and chief of psychiatry at Beth Israel Hospital in Boston. Bibring made the case that pregnancy was central to an understanding of women's inner lives: "For a number of years an interest has been growing increasingly in the psychological aspects of pregnancy," she wrote in 1959. "The significance of this condition for the individual and for the family constellation cannot be overestimated by any student of human behavior." Despite its importance, most of our ideas on the subject come from "the folklore and old wives' tales concerning the emotional instability and unpredictability of pregnant women," she observed in another article: "We all know the proverbial story of the woman who rouses her husband in the middle of the night to insist that he must find her fresh strawberries or peaches in the middle of January."

Bibring aimed to look beyond such stock story lines to discover what went on in women's psyches during pregnancy. She initiated a large, long-term study of pregnant women at Beth Israel, out of which emerged some deeply intriguing ideas. Pregnancy is a developmental phase, akin to adolescence, in which the psyche changes along with the body, Bibring proposed. It is a "normal crisis," a "point of no return between one phase and the next," which must be navigated successfully to reach the next level of maturity. During these fraught nine months, old conflicts emerge, old solutions break down, and new roles—indeed, a new self—must be constructed. Pregnancy is not a passive period of waiting, nor a pathological state of raging hormones. It's a phase unto itself, vital, dynamic, and creative, though at times uncomfortably unruly. And its outcome helps determine how well we handle the next phase: just as the course of adolescence can influence young adulthood, pregnancy sets the tone for motherhood. "There is no turning back" from this developmental challenge, Bibring concludes; pregnancy is a "testing ground of psychological health."

A few days after reading Bibring, I'm still thinking about her ideas as I lace up a pair of sneakers. I'm preparing to exercise, something I haven't done much since I was—well, a teenager. When my feelings overwhelmed me, as they often did in those days, I found some peace of mind in running. Fresh from a fight with my mother or some misunderstanding with a boyfriend, tears still wet on my face, I would pull on my sneakers and head out the door, on my way to the elementary school down the road. Around the perimeter of empty playing fields, I ran as hard as I could, liking the jarring feeling of my feet striking the ground, glad that my breathing was too loud to hear the recent argument replaying in my head. Finally after three-quarters of an hour I would slow, wiping the sweat from my face with a forearm, checking gingerly on my anger and self-pity. Almost always, I'd find that I'd outrun them.

Once the dramas of adolescence passed I gave up the running habit, but now I consider taking it up again. I know from my reading that exercise can alleviate depression during pregnancy, as it does at other times in women's lives. Research suggests that it may also reduce the risk of preeclampsia and gestational diabetes, and help manage musculoskeletal problems like low back pain.

More surprisingly, exercise may also make the fetus healthier. Linda May is an assistant professor of anatomy at Kansas City University of Medicine and Biosciences. Using a device called a biomagnetometer, which measures the magnetic fields produced by electrical activity in the heart, May conducted a study of the effects of maternal exercise on the fetus. Comparing sedentary pregnant women to those who engage in moderate-intensity aerobic exercise for at least thirty minutes three times a week, May found that fetuses show the same beneficial effects of cardiovascular training as do their physically active mothers: their heart rates are significantly lower, and their heart-rate variability is greater, than those of fetuses of mothers who don't exercise. May is now investigating whether exercise during pregnancy could be used as a very early intervention to improve children's health.

Exercise may even make the fetus smarter. In a recent widely praised book, *Intelligence and How to Get It*, University of Michigan psychologist Richard Nisbett argues that IQ is more malleable than once believed. One of his principal pieces of advice for raising smarter kids: work out while you're pregnant. Women who exercise while expecting tend to have larger babies who grow up to be smarter adults, Nisbett writes, perhaps because their brains are bigger.

And so I find myself at a gym a few blocks from my apartment, feeling acutely aware of my six-months-pregnant form as dozens of toned bodies bob and gyrate around me. I have my doubts, too, about the hulking treadmill I'm about to step on: can this piece of equipment give me what I once got from fresh air and hard-packed dirt? But as the treadmill starts up its steady hum, I discover that I like the high-tech machine and the pleasantly pliable reality it offers. I punch away at its buttons, making my jog faster and then slower, steeper and then flatter. As the manufactured miles slip past, as I scale the tilted-treadmill hills and descend its valleys, I find that for the moment I've left my gloom behind.

If pregnancy provokes a "normal crisis" for many women, for others it can accompany a crisis that's far less benign. Research shows that mood disorders are no more common among pregnant women than among their nonpregnant peers. But it's during a woman's childbearing years—roughly twenty to forty—that she is at the highest risk of depression or anxiety, and contrary to what doctors once believed, pregnancy confers no protection. Psychiatrists estimate that about 20 percent of pregnant women experience mood or anxiety disorders and about 10 percent develop major depression, though many of these cases go undetected. That may be in part because the symptoms of psychiatric illness overlap with those of pregnancy: changes in sleeping patterns, eating habits, and energy levels.

Of all the psychological problems that occur in pregnancy, we know the most about depression. Even so, antenatal depression has received far less attention than the postpartum kind, highlighted by celebrity sufferers like Brooke Shields, or even the rare phenomenon of postpartum psychosis, which led Andrea Yates to drown her five children in a widely publicized case in 2001. Yet depression during pregnancy may be more common than depression that sets in after birth. Researchers in Avon, England, followed more than fourteen thousand women through pregnancy and childbirth, tracking their symptoms of depression at each stage. "More mothers moved above the threshold for depression between eighteen weeks and thirty-two weeks of pregnancy than between thirty-two weeks of pregnancy and eight weeks postpartum," reported the study, published in the *British Medical Journal* in 2001. "Research and clinical efforts need to be moved toward understanding, recognizing, and treating antenatal depression." There's good reason to do so: about half of women depressed during pregnancy will be depressed after birth as well, making it, as one researcher puts it, "the single most predictive factor for postpartum depression."

Once depression is diagnosed, the matter grows only more complicated. Today, treatment for depression usually entails the prescription of antidepressant medication, often a selective serotonin reuptake inhibitor, or SSRI. But the use of SSRIs during pregnancy has been associated in some studies with an increase in breathing problems in the newborn. "Pregnant depressed women are really on the horns of a dilemma," says Shari Lusskin, a professor of psychiatry and obstetrics and gynecology at New York University. "Many of them don't want to take medication, but without it they really struggle." Research shows that women who go off antidepressants when they become pregnant run a high risk of relapse. A 2006 study published in the *Journal of the American Medical Association* followed 201 pregnant women with a history of major depression. Of the women who continued to take their medication, about a quarter became depressed during pregnancy; among the women who stopped taking the drugs, more than two-thirds experienced a relapse.

And depression itself, it's now recognized, is associated with problems during pregnancy. Pregnant women who are depressed are more likely to have their babies prematurely. A 2008 study by researchers at the Kaiser Permanente Medical Center in Oakland, California, found that women with mild symptoms of depression are 60 percent more likely to deliver early; women who are severely depressed have twice the risk of premature birth. The babies of depressed women are also more likely to have low birth weight. These complications may come about because depressed women don't take good care of themselves: they may eat poorly, smoke or drink alcohol, or fail to get prenatal care. Or depression itself may shift the biochemical balance in a woman's body in a manner that produces early delivery and low birth weight. For example: the stress hormone cortisol, which is often elevated in individuals with depression, may cross the placenta to directly affect fetal development, or it may affect the functioning of the pregnant woman's blood vessels, thereby reducing the oxygen and nutrients that reach the fetus. "There's no risk-free option," says Lusskin. "But in most cases, the risks of treatment, including medication, are less serious than the risks of untreated depression."

Increased rates of premature delivery and low birth weight among babies born to depressed pregnant women have been firmly established by research. Now scientists are exploring a startling but still speculative notion: that a pregnant woman's emotional state can influence the fetus's developing brain and nervous system, potentially shaping the way the offspring will experience and manage its own emotions—a kind of maternal impressions redux.

I first encounter this idea at a meeting of a group of fetal development experts, the Perinatal Brain and Behavior Network, held in Washington, D.C. One of the presenters is Catherine Monk, a professor of psychiatry at Columbia University. She stands at the front of the room, dressed in a chic black suit, her hair twisted into a sleek chignon.

"New research indicates that even before birth, women's moods may affect child development," Monk announces. "Our research is interested in the following questions: Can maternal mood be transmitted to the fetus? If so, what is the mode of transmission? And how do such moods affect fetal development?

"These are new questions to be asking," she adds. "We're still figuring out how to get fetuses to answer."

In fact, Monk and her colleagues have gone some distance toward putting the fetus on the couch. When we're both back in New York, I get in touch with Monk, and she invites me to observe one of the studies she has underway. On a morning in early December, I arrive at Presbyterian Hospital in upper Manhattan. I'm greeted by Andrea, a friendly young graduate student dressed in a white lab coat. She shows me into a brightly lit control room, humming with equipment, where a technician named Michael is already tapping away at a computer. Visible through a one-way mirror is an examining room, quiet and dim, with wan winter light filtering in through a shade. This is where Andrea now welcomes today's subject, a woman three weeks shy of her due date.

For this study, Catherine Monk is comparing four different groups of pregnant women: those who are depressed and receiving antidepressant medication; those who are depressed and receiving therapy; those who are depressed and not in treatment; and those with normal mood, included as an experimental control. The woman who's just entered the examining room belongs to the untreated depressed group, a fact apparent even to me. She looks at the floor, avoiding eye contact, and when she introduces herself her voice barely rises above a whisper. As Andrea settles her in an armchair, the woman rests her head against the back of the chair and gazes listlessly out the window. She moves little and says nothing as Andrea slips a blood-pressure monitoring device over her middle finger and applies two electrodes just below her collarbone. When Andrea gently lifts the woman's shirt, exposing her abdomen, I'm startled to see the fetus move under her skin, a gentle ripple.

Watching from inside the control room, I can almost feel the lead weight of her sadness, how the mere act of existing seems to take more energy than she can muster. I'm reminded of something another researcher told me: that during ultrasound examinations, a depressed woman will often show no interest in looking at her fetus on the screen. I have every reason to feel empathetic, but to my chagrin I find that I feel repelled. Her drawn face is jarring above the lush curve of her belly, and the deadness of her affect seems painfully at odds with the life moving inside her. For the first time, I begin to understand why the notion of depression during pregnancy arouses such discomfort.

As Andrea applies gel to the woman's skin, she asks a few casual questions: Is this her first baby? Does she know the baby's sex?

"It's a boy," the woman says, and smiles for the first time, faintly. Andrea places a third electrode on the woman's belly and moves it around, searching. Soon, the familiar sound of a fetal heartbeat fills the control room, and the number of its beats per minute appears on a monitor. Michael taps a few keys, and five lines stretch across his screen: they graph the electrical activity of the woman's heart, her blood pressure, and her breathing; the two lines at the bottom display the fetus's heart rate and its movement.

Andrea returns to the control room, and the experiment begins. The woman is taking a standard psychological test known as the Stroop task, in which the participant must identify the names of colors. The tricky part is that the letters themselves appear in a color other than the one the word spells out, producing momentary confusion. I watch as the woman pushes buttons on a handheld keypad in response to the words flashing on the screen in front of her: "Green," reads the word on the monitor, but the letters are yellow; "Blue," reads the next word, written in red. The screen posts the woman's results after each round. "Incorrect," it reads. "Incorrect." "Incorrect."

"She's not doing so well," Michael observes.

"She said she didn't want to practice first," Andrea murmurs.

"Not giving it the old college try," Michael replies dryly.

In silence we watch the woman struggle on for a while; then Michael leans into a microphone, hooked up to a speaker in the examining room.

"Work faster, please," he intones.

This is a standard part of the experiment, intended to increase the stress level of the subject. Sure enough, the lines graphing the woman's heart rate, blood pressure, and respiration start to spike. Michael waits a beat, then repeats his command: "Work faster, please." Now I notice that the fetus's heart rate is rising, too. All the while the words keep flashing on and off the screen: "Incorrect." "Incorrect."

At last it's over. Andrea leans into the examining room. "You can rest quietly for a few minutes now," she says, her voice gentle. The woman doesn't answer: she just allows her head to fall back against the chair again, and closes her eyes.

What Catherine Monk has learned from sessions like these is that the fetuses of depressed and anxious women are especially reactive. When given the Stroop task, all the pregnant women in Monk's experiments show increases in heart rate, blood pressure, and respiration. But only the fetuses of women who are depressed, or who have anxious personality styles, display heart-rate increases of their own. "That difference suggests that these fetuses are already more sensitive to stress," Monk tells me. "Perhaps that's because of a genetic predisposition, inherited from the parents. Or it could be because the fetuses' nervous systems are already being shaped by their mothers' emotional states." Women's heart rate and blood pressure, or their levels of stress hormones, could affect the intrauterine milieu over the nine months of gestation, Monk explains, influencing what is an individual's first environment and thereby affecting its development.

The differences Monk has found among fetuses appear to persist after birth: fetuses who were more reactive to stress become babies who are more reactive to stress. And because basic physiological patterns, like heart rate, are associated with more general differences in temperament, Monk says, "it may be that the roots of temperamental variation go back to the womb." She points to studies by other researchers finding that the newborns of depressed mothers are more irritable and hard to soothe, have more problems sleeping, and have higher levels of the stress hormone cortisol in their blood. Later on in children's lives, maternal depression and anxiety during pregnancy is associated with higher rates of impulsivity, hyperactivity, and emotional and behavioral problems.

Beyond temperament, it could even be the case that conditions in the womb influence later susceptibility to mental illness. "We know that some people have genetic predispositions to conditions like depression and anxiety," Monk points out. "And we know that being raised by a parent with mental illness can increase the risk of mental illness in the offspring. It may be that the intrauterine environment is a third pathway by which mental illness is passed down in families." This kind of research, says Monk, "is pushing back the starting line for when we become who we are."

Listening to Monk, I find myself thinking of patterns of emotion in my own family. My episodes of Mommy Despair, I suspect, have their roots in a familial tendency to high-strung nervousness; my mother and I are both wired in a way that makes us jump at sudden noises and go white-knuckled on the highway. Anxiety has been passed from mother to child like an old debt, a cracked heirloom no one wants. Until now, I'd blamed an unlucky piece of DNA, or a parenting style that pointed out every potential danger. Now, I wonder if the handoff began in the womb, and if there's anything I can do to help my second son evade this legacy.

A few years ago, I did find a way to quiet the long-playing loop of warnings and alarms in my head: meditation. I even joined a meditation group that met each Thursday evening at the home of its leader, a Zen teacher and psychotherapist named Anna. In Anna's Upper East Side apartment, filled with books and strewn with Oriental rugs, I sat on a round cushion and learned to focus on my breath, to follow its movement in and out of my lungs instead of the antic whirl of ideas in my head. After Teddy was born my attendance dwindled, but when I show up again one Thursday night in December, the other members accept my presence with a smile and a nod.

We assemble in a circle on Anna's living-room rug, and she strikes a bell to begin the first sitting. Even before its reverberations have died away I can sense my mind settling down, and feel my fetus pausing in his synchronized swimming routine. Another chime of the bell, another sitting; by the time Anna begins to speak in her soft, steady voice, I feel at once alert and at ease. Tonight, she asks us to contemplate a koan, a riddle with no solution: "Not one, not two." Our minds are not separate from our bodies, Anna explains; people are not truly separate from each other, or from the matter that makes up the rest of the universe.

Once I would have inwardly dismissed such talk as so much New Age mystification, the price of admission to the mental training I did value. But tonight it strikes my ear differently; it makes sense of the paradoxical condition in which all pregnant women find themselves. As the meeting ends and I stand up, stretching my legs, I can feel my fetus resume his aquatic acrobatics: he and I are not one, and not two.

In addition to investigations like Catherine Monk's, there is another line of research now exploring the potential prenatal origins of mental illness, in which the proposed mechanism is not maternal mood but experiences of extreme stress or malnutrition. War can be one of these extreme stressors, as I'd already learned from the research of New York University psychiatrist Dolores Malaspina. She found that women who were pregnant during the Arab-Israeli Six Day War gave birth to offspring who were significantly more likely to develop schizophrenia as young adults.

A different kind of trauma, in another part of the world, also appears to have led to elevated rates of schizophrenia. In 1959, the Chinese Communist leader Mao Zedong announced a "Great Leap Forward." China, he declared, would transform itself from a land of peasants to a modern nation full of factories. He commanded tens of millions of farmers to begin making steel instead of growing rice. Within just a few years, Mao promised, China's steel output would surpass Great Britain's. What happened instead was the worst famine in world history, killing some thirty million people and causing unimaginable suffering among the survivors. The widespread starvation also affected fetuses in utero, in an unexpected way: they grew up to develop schizophrenia at a higher rate than individuals conceived before or after the famine. Neuroscientist David St. Clair, of the University of Aberdeen in

Scotland, examined thirty years of psychiatric case records from the Wuhu region of Anhui, where residents experienced severe malnutrition during the Great Leap Forward. His findings, published in the *Journal of the American Medical Association* in 2005, showed that individuals born to women during the famine were twice as likely to develop schizophrenia as those born at other times.

Schizophrenia is a complex disorder with many potential causes, but work by St. Clair and others indicates that severe maternal malnutrition can be a contributing factor in the disease's development. In particular, research is pointing in the direction of folate: it may be that a lack of this critical nutrient during pregnancy produces new mutations in DNA, prevents the effective repair of DNA, or disrupts the way genes are expressed, leading to a higher incidence of schizophrenia. The findings from the famine caused by the Great Leap Forward are buttressed by similar studies of individuals conceived during the Dutch Hunger Winter of 1944. Earlier research by Ezra Susser, a professor of epidemiology and psychiatry at Columbia University, had reported results almost identical to those from China: individuals in utero at the time of the Nazis' siege of the Netherlands had double the risk of developing schizophrenia. "The fact that such similar results were found in two very different populations indicates that we're onto something real here," Susser tells me.

Startling though it might seem to trace the roots of mental illness back to the womb, the notion comes as no surprise to Susser. He is the son of Mervyn Susser and Zena Stein, pioneering epidemiologists who decades ago led the original investigations into the effects of the Dutch Hunger Winter. "Growing up, I heard about the Hunger Winter every day. When I was a teenager I did clerical work on the studies," says Susser. "So when researchers first proposed prenatal origins for schizophrenia, it didn't strike me as odd. The notion that conditions we experience as adults might have their start before birth seemed like the most natural thing in the world."

Over the long term, says Susser, research like his could lead to a more complete understanding of schizophrenia—perhaps even a strategy for preventing this rare but devastating disease. In the meantime, we can take a step toward addressing a disorder that is all too common in pregnancy: depression. Obstetricians could screen pregnant women for depression, just as they do for a host of other conditions, and refer those who need psychological help to treatment. Maternal depression occurs at least as often as pregnancy-induced hypertension, and five to ten times as often as gestational diabetes, both of which are checked in the course of routine prenatal care. But few OBs check for depression, says the reproductive psychiatrist Shari Lusskin. "Obstetricians will tell you that they're too busy, that they aren't equipped to handle patients' psychological issues," Lusskin tells me. "Or they believe that their parents don't experience depression, that pregnancy protects them from negative moods."

There are now several instruments available for screening pregnant women for depression; they're quick enough to be completed by patients in the waiting room, while we would otherwise be flipping through last month's magazines and checking out each other's bellies. I take a look at one of them: the Pregnancy Depression Scale, a simple, seven-question test that identifies depression in pregnant women with an impressive degree of accuracy. Reading over the test, I find myself answering its questions in my head. Do I feel sad, hopeless, helpless, worthless? Am I full of self-reproach, do I feel that I've let people down? Do I ruminate over past errors? Do I experience listlessness, indecision, and vacillation? Although it's intended to be scored by a doctor, at the end of the test I do a quick tally of my responses. I don't have depression, the score key reports, but I may be at risk: "Monitor closely."

In all my visits to my obstetrician, I realize, she has never asked me about my emotional state, and I've never volunteered to talk about it; together we colluded in the fiction that everything was fine. Along with all the practical objections they raise, I wonder if there isn't another, more human reason that obstetricians don't screen their patients for depression: they need pregnant women to be happy, too.

It's when I catch myself humming Christmas songs that I realize I've begun to feel better. I'm going to the gym twice a week now, and stopping to focus on my breath whenever anxious thoughts start to pick up speed. Exercise and meditation are clearly working; still, I wonder if there's anything else I should try—like therapy. Until recently, there was little research looking specifically at the effectiveness of therapy for pregnant women. Margaret Spinelli, director of the Women's Program in Psychiatry at Columbia University, is working to change that. Spinelli has adapted a type of treatment called interpersonal therapy to the distinctive needs of pregnant women. "Interpersonal therapy focuses on individuals' relationships and on the roles they play with family, friends, and community members," Spinelli tells me. "Relationships, both past and present, assume a special importance when a woman is pregnant." Women enrolled in the program Spinelli has designed meet weekly with a therapist for three months; together, they work on issues like improving communication, clarifying expectations, and mourning the past loss of loved ones, as well as issues particular to pregnancy, such as resolving ambivalence or apprehension about becoming a mother and addressing the changes a child brings to the parents' relationship.

In 2003, Spinelli published the first controlled clinical treatment trial for therapy during pregnancy. The study, which compared the effects of interpersonal therapy to a parenting-education program on a group of fifty depressed pregnant women, found that those in the therapy group showed significant improvement, with 60 percent achieving recovery. (Only 15 percent of the women in the education group recovered from their depression.) Spinelli is now testing the program on a larger group of subjects. Reading the manual she wrote for therapists treating pregnant women, I'm struck by the echoes I hear of the work of Grete Bibring. "Pregnancy, a time of developmental upheaval, is similar to the onset of adolescence," Spinelli writes. "The woman's self-concept is reorganized as she adapts to her new role as mother." It is, she adds, "among the most common and complete transformations in human experience."

On a morning in late December, I slip into the hush of a suite of offices in Upper Manhattan. I'm here to see Catherine Monk, who practices therapy with pregnant women in addition to conducting research with them. Monk greets me in the waiting room; today her hair is soft around her shoulders, and she's wearing a fuzzy chartreuse sweater—the sleeve of which, she shows me with a laugh, has been chewed by her new puppy. She shows me to her office and goes to fetch me a glass of water. Taking a seat, I look around. Her bookshelves are heavy with psychiatric and obstetric texts—and, rather charmingly, a copy of *Pregnancy for Dummies*. On her wall hangs a framed poster from a museum exhibit, an abstract painting of blue and green and red and yellow. I imagine her pregnant patients gazing at this picture while they talk, tracing its languid brushstrokes with their eyes as they tell Monk what it's like to have another life growing inside them.

Monk returns, and I begin our interview by asking her what's wrong with the way we think about emotions during pregnancy. "We treat pregnancy as if it's a medical condition," she replies. "Just as we've medicalized depression—thinking of it only in terms of neurotransmitters—we've medicalized pregnancy, made it a matter of hormones, when really it is a profound emotional and psychological experience." Seconding her Columbia colleague Margaret Spinelli, Monk points out that pregnancy unfolds not just in obstetricians' offices but in a web of relationships with partners, relatives, and friends. Perhaps most important of all, Monk says, is the relationship between the pregnant woman and the fetus. "Women are often very attuned to the particular characteristics of their fetuses during pregnancy. They tell me, 'This baby is going to be feisty—her kicks are so strong,' or 'This baby is a night owl—he always moves around when I'm trying to go to sleep,'" she says. "Women are starting to create representations of their babies while they're still pregnant, starting to build relationships with them even before birth."

In therapy, Monk says, she tries to "bring the baby into the room," helping the woman to examine her feelings about the fetus, and to identify which patterns of relating, or "templates," she's applying to her child-to-be. "We all have templates for our relationships, and these templates are about our expectations of the other person, how reliable or unreliable they are, how worthy we are of their love," she says. "If an unhelpful template is being applied to the fetus—taken from, say, the woman's difficult relationship with her own mother—we try to find a different template for her to access. Maybe she had a close relationship with a sister or an aunt, for example, that she can bring into play." Remarking on the connections between her laboratory research and her therapy practice, Monk notes that there's another way in which the baby is "in the room": if maternal mood indeed affects the fetus, than therapy for the mother is therapy for the fetus, too. "By helping a pregnant woman to better regulate her moods, we could potentially influence the course of fetal development," she says.

Another important and often overlooked aspect of pregnancy, Monk says, is the seismic change it can wreak on a woman's identity. "We're used to thinking of adolescence as a time when our bodies are changing, when our emotions are unruly—well, pregnancy is very similar," Monk tells me as our interview nears its end. "It's a very disorderly time, when a lot of things are in flux. But that fluidity also opens up new opportunities for positive change." She pauses and meets my eyes; suddenly I have the disconcerting feeling that my own complicated emotions are visible to her penetrating gaze. "You have to let yourself fall apart, and then put the pieces back together in a different way."

After saying my goodbyes, I stand for a moment inside the tinted glass doors of Monk's office building, pulling on my gloves and wrapping a scarf around my neck. It's a few days before Christmas, and the scene at the corner of West 168th Street and St. Nicholas Avenue is full of disorderly life: throngs of people laden with shopping bags, tinsel strung from the streetlights, a guy dressed like Santa selling newspapers outside the subway. I take a deep breath, and head out into the brilliant winter sun.

SEVEN MONTHS

Now that I'm in my seventh month, my obstetric appointments are scheduled two weeks apart, and it sometimes feels as if I spend most of my time walking to and from my doctor's visits. Not that I mind. My obstetrician's office is in SoHo, that layered New York neighborhood where camera-toting tourists, bag-laden shoppers, actors, models, and a last few painters and gallery owners all elbow past one another on the crowded sidewalks. Making my way, slowly, over Prince Street to my OB's office on Broadway, I let my eyes fall for a moment on each individual I pass, and I find myself struck by a sudden thought: every person on this street was once a fetus. As epiphanies go it's pretty obvious, and yet the longer I contemplate it the stranger it seems. Our origins story— that each of us spent nine months inside a woman's uterus—is as fanciful, as implausible on its face as any primitive people's creation myth. We know it's true, but we don't quite believe it.

I was reminded of just what a just-so story our fetal beginnings can seem when I tried it out on Teddy for the first time. After seven months of not noticing—or studiously ignoring—

169

my pregnancy, Teddy looked up from his blocks one morning last week and asked, "Mommy, why is your tummy so big?" Recognizing an official teachable moment in the making, I took a breath and launched into an explanation: how there was a baby growing in my belly; how, after nine months, the baby would come out; how Teddy, too, had come into the world this way. I tried not to be distracted by the look on his face, a deepening expression of disbelief and even suspicion. "And that's how every single person got his start!" I concluded, with what I hoped was an air of incontrovertible authority. He looked thoroughly dubious now, and for a moment I flashed forward ten years, to a vision of Teddy as a budding teenager. He turned back to his elaborately ramparted castle. "That's silly, Mommy," was all he said.

Now, looking around this downtown block, I understand the familiar facts of life as if for the first time. Those burly guys unloading boxes outside a bodega, these spindly girls emerging from a boutique: former fetuses all. In place of their thick hides and sharp collarbones, I see the translucent skin of a fetus, the lacy tracing of veins, vertebrae delicate as a fossil. The vision makes me feel tenderly toward each of them, once so vulnerable and new to the world. Stopping at the corner to catch my breath, I suppress a smile at my sentimentality. Mine is an emotional reaction, born of my current condition and its attendant preoccupations. But it's not so different from a perceptual shift now underway in scientific laboratories all over the world. Researchers are beginning to look at adults from a new angle: as former fetuses, organisms shaped by prenatal experience.

It's an idea as self-evident, and as startling, as my episode of double vision on a SoHo street is to me. Clearly, what comes first has to affect what comes after; that's apparent enough in the case of an underweight newborn whose mother was malnourished during pregnancy, or the child of an alcoholic who bears the stigmata of fetal alcohol syndrome. David Barker, the British physician who was an early advocate of the notion that many of our physiological characteristics originate during gestation, says that when he shares his radical insight with ordinary mothers and grandmothers, "they often look at me pityingly, as if I'm a bit thick to have just come to this realization." And yet, carried to its logical conclusion, the idea is positively mind-bending. Could it really be the case that our child and even our adult selves are shaped by what happened to us before birth?

Take Barker's original hypothesis, that heart disease has its roots in the prenatal period. On one level, it seems plausible enough. The heart is made during fetal life, after all, and we emerge from the womb with its four chambers already pumping away. Couldn't some flaw or weakness in its construction show up decades later— like a car with a part poorly welded in the factory that breaks down years after its purchase? And yet when Barker introduced his idea the reaction was scathing: he was roundly mocked by members of his profession, and some who heard it even walked out of his lectures in disgust. Heart disease is the product of genetics, he was informed, and of lifestyle factors in middle age, such as smoking and lack of exercise. When Barker was at last grudgingly recognized with an invitation to a conference of American scientists, no name tag could be found for him (the organizers finally gave him one that read "Technician"), and he lectured late in the evening to an almost-empty room.

Two decades later, the medical establishment has come around to the possibility that Barker's crazy idea is correct. Genes and lifestyle clearly account for much of the risk of heart disease, but prenatal factors may also play a part. And the notion that fetal experience might have lasting effects is now reaching past heart disease into other conditions—past disease itself, in fact. The original name for this field was the "fetal origins of adult disease," or FOAD. That acronym has been swapped for another: DOHaD, or the "developmental origins of health and disease." The first change—"developmental" for "fetal"—reflects the recognition that some important influences occur soon *after* birth: what and how much a newborn is fed, for example. (Since this book is concerned with the nine months of pregnancy, I've chosen to use the term "fetal origins.") The second change—"health and disease" for simply "disease"—marks the realization that strength and vitality, as well as frailty and vulnerability, may also start very early in life.

This simple, and in Barker's self-deprecating telling, rather obvious idea has already had an outsized impact, spawning international conferences, ambitious research programs, and thousands upon thousands of scientific publications. In this rapidly expanding literature, one can now find reference to the fetal origins of cancer, asthma, obesity, diabetes, mental illness—even of conditions usually associated with old age, such as arthritis, osteoporosis, and cognitive decline. This research is still in its early stages. Much of it involves animals rather than humans, or comes with methodological limitations: studies that ask women years later to recall what they ate during pregnancy, or that draw conclusions from relatively small numbers of subjects. As one might imagine, research that seeks to discover connections between fetal life and adult life is extraordinarily difficult to execute; to do it right will take decades. In the meantime, we have no shortage of deeply intriguing findings, results that suggest startling new answers to some very familiar questions.

* * *

Begin with one of humankind's oldest ideas: the notion of inherited guilt, in which flaws and failings are handed down from one generation to another. The ancient Greeks told of family curses, such as the one inflicted on the House of Thebes: because King Laius had offended the gods, Laius's son Oedipus was fated to kill him. The authors of the Old Testament book of Exodus warned of "visiting the iniquity of the fathers upon the children, and upon the children's children, unto the third and to the fourth generation." And centuries of folklore blamed individuals' deviancy on their families' "bad blood."

In the nineteenth and early twentieth centuries, the notion of inherited guilt was placed on a purportedly scientific basis by the adherents of eugenics, who believed they could improve the human race by preventing the "unfit" from reproducing. They wrote up elaborate case studies of individuals they deemed defective, and carefully catalogued their extended families, using these pedigree studies to demonstrate how promiscuity, laziness, criminality, pauperism, and feeble-mindedness ran in the family. They christened these clans with scornful pseudonyms, like the Zeroes or the Nollys or the Jukes ("juke" being a slang term for chickens who don't lay their eggs in nests, but deposit them wherever the whim strikes). And they toted up how much these families cost society in relief payments, medical care, and jailhouse room and board. The total, the eugenicists noted indignantly, came to millions of dollars. One of the most notorious of these clans was the so-called Kallikak family.

The Kallikaks were brought to public attention by Henry Goddard, director of the research laboratory of the Vineland Training School for Feeble-Minded Boys and Girls in New Jersey.

Goddard noticed that a large number of the school's students were related to each other; his curiosity piqued, he traced the family line back six generations. A single feeble-minded man, he concluded, was the forbear of hundreds of defectives. In 1912, Goddard published a detailed account of the clan, *The Kallikak Family: A Study in the Heredity of Feeble-Mindedness*. The book became a bestseller, and the Kallikak family became infamous, cited in textbooks and encyclopedias as a cautionary tale about unchecked reproduction among the unfit. The story of the Kallikak family was used as an argument for the segregation, the sterilization, and even the extermination of such people; in translation, the book was popular reading among members of the Nazi party in Germany.

Goddard and his followers firmly believed that a "hereditary taint" was to blame for the Kallikaks' deficiencies: "That we are dealing with a problem of true heredity, no one can doubt," he wrote. The members of the Kallikak family "are wayward, they get into all sorts of troubles and difficulties, sexually and otherwise, and yet we have been accustomed to account for their defects on the basis of viciousness, environment, or ignorance." These explanations, he insisted, are mistaken. "The question is, 'How do we account for this kind of individual?'" Goddard asked. "The answer is, in a word, 'heredity,'—bad stock."

Almost a century later, a doctor named Robert Karp wondered otherwise. Karp, a pediatrician at the State University of New York Downstate Medical Center, works with many patients with fetal alcohol syndrome, and he thought he recognized some of its signs in what he read about the Kallikaks. One day in 1994, he traveled to the Vineland Training School (now reinvented as a human-services organization called Elwyn New Jersey) and began combing through its records.

He soon came across the file of Pauline, whose chart noted that she was mentally retarded and had stunted growth and a small head circumference—all well-recognized symptoms of fetal alcohol syndrome. Her photo clinched it: it showed a gap-toothed young woman, smiling broadly for the camera, with a white bow pinned to her sandy hair. "It was clear to see that she was affected by fetal alcohol syndrome: she had the flattened features, the smooth upper lip, and the slitlike eyes of someone with FAS," Karp tells me when I reach him at his office in Brooklyn. This suggested an explanation for the characteristics of Pauline and the other Kallikaks that was entirely different from Goddard's. "Of course, alcoholism itself has a strong genetic component, but many of the attributes Goddard assumed were hereditary—the mental retardation or 'feeblemindedness,' the slowed motor movements that he called laziness, perhaps even the lack of impulse control that led to uninhibited sexual behavior—may well have stemmed from these individuals' exposure to alcohol in the womb."

Karp's analysis of the records and photographs he found in the Vineland files, published in the *Archives of Pediatrics and Adolescent Medicine* in 1995, demonstrated that many students at the school likely suffered the ravages of fetal alcohol syndrome. He is not the only contemporary researcher to theorize that the Kallikaks and other families maligned by the eugenicists were actually shaped by a variety of adverse prenatal influences like alcoholism, malnutrition, and venereal disease, not to mention poor conditions after birth. But the emotional force of Goddard's story—the image of one generation of feeble-minded deviants heedlessly begetting another—resonated for decades, obscuring the influence of prenatal and postnatal environments. The tale of the Kallikaks, biologist and historian of science Stephen Jay Gould has noted, became "a primal myth of the eugenics movement."

The theories and methods of the eugenicists have long been soundly repudiated. But some of their assumptions about the causes of dysfunction linger on in the way we sometimes think and talk about today's poor—as irredeemable, incorrigible, doomed to repeat the destructive patterns of their forbears. Such assumptions have important implications for social policy: why try to help the disadvantaged if they are constitutionally incapable of being helped? Pushing back against such fatalism is decades of psychological and sociological research demonstrating the powerfully warping effects of an impoverished childhood environment. More recently, neuroscience has weighed in: researchers at Cornell University, the University of Pennsylvania, and the University of California, Berkeley, among others, have demonstrated that the stress of childhood poverty can interfere with the development of memory, problem-solving, and language skills. Now still another factor is coming to the fore: the realization that poor children may enter the world already disadvantaged by adversity experienced before birth.

Consider: Poor pregnant women are more likely to drink heavily, smoke, and use drugs than affluent pregnant women. Poor pregnant women are more apt to have inadequate diets, including specific nutritional deficiencies, and are less likely to take vitamin supplements. Poor pregnant women are exposed to more environmental toxins, including second-hand smoke, industrial emissions, pesticides, and lead. Poor pregnant women encounter more daily stressors, and are more likely to suffer from depression and anxiety. Poor pregnant women experience more trauma, and have fewer resources to help them cope with such experiences. Poor pregnant women are less likely to have health insurance and to get adequate prenatal care. Poor pregnant women are at higher risk of having premature and low birth weight babies.

The effect of such adverse influences, especially cumulatively, could well have an effect on the intellectual abilities and physical robustness of their offspring, and perhaps even on the propensity to use drugs or engage in criminal activity. (Human and animal studies suggest that prenatal exposure to nicotine, alcohol, or cocaine can change the fetal brain in ways that make offspring more likely to experiment with or become addicted to such drugs as adolescents or adults. And several epidemiological studies have linked early lead exposure to higher rates of juvenile delinquency and adult crime.) Given the odds stacked against poor women and their fetuses, the most effective antipoverty program might be one that starts before birth.

Take another, often controversial, question about the origin of a human behavior: why are some men gay? Sigmund Freud supplied one influential answer: growing up in a particular family constellation, made up of a domineering and "seductive" mother and a weak or absent father, could turn a boy into a homosexual adult. Promoted by psychoanalysis, the notion that certain styles of parenting produced homosexual offspring persisted through much of the twentieth century. In recent decades, that idea has been supplanted by excitement about research into the genetic basis of gayness. In 1993, Dean Hamer, a geneticist at the National Institutes of Health, announced that a specific region of the X chromosome was found more often in gay than in straight men (although other scientists have failed to replicate Hamer's finding). Twin studies have demonstrated that identical twin brothers of gay men are more likely to be gay than are fraternal twin brothers of gay men, suggesting some genetic contribution to homosexuality. As with all complex human behaviors, there are almost certainly many factors that contribute to the shaping of an individual's sexual orientation. Now, we may be able to add fetal experience to the list.

The first clue pointing in the direction of prenatal influences was the odd fact that men with older brothers are more likely to be homosexual. This "fraternal birth order effect" has now been documented in more than a dozen studies: the more older brothers a man has, the better the chance that he will be gay. The pattern can even be found in the data of the pioneering sex researcher Alfred Kinsey, though Kinsey himself didn't notice it. Freud did take note of the phenomenon, and once again had a theory to explain it. "Observation has directed my attention to several cases in which during early childhood feelings of jealousy derived from the mother-complex and of very great intensity arose against rivals, usually older brothers," Freud wrote in a 1923 paper titled *Certain Neurotic Mechanisms in Jealousy, Paranoia, and Homosexuality*. "This jealousy led to an exceedingly hostile aggressive attitude against brothers"; eventually, "these feelings yielded to repression and to a transformation, so that the rivals of the earlier period became the first homosexual love-objects."

Anthony Bogaert, a professor of psychology at Brock University in Ontario, Canada, offers a different explanation. Along with Ray Blanchard, a professor of psychiatry at the University of Toronto, he has suggested that during pregnancy, the mother's immune system manufactures antibodies directed at particular proteins produced by male fetuses. When she becomes pregnant with another male fetus, those antibodies affect his developing brain, predisposing him to homosexuality. "The hypothesis doesn't explain all homosexuality, of course, but we estimate that about a third of gay men are gay because their mothers had sons before them," Bogaert tells me. Evidence for the "maternal immunization hypothesis" was bolstered in 2006 when the *Proceedings of the National Academies of Science* published Bogaert's study of nearly a thousand men. He found that biological older brothers made a man more likely to be gay, whether he'd grown up with them or not; older stepbrothers raised in the same household had no effect on sexual orientation.

"These data, by elimination, strengthen the notion that the common denominator between biological brothers—the mother—provides a prenatal environment that fosters homosexuality in her younger sons," noted a commentary that accompanied Bogaert's study. The commentary's authors couldn't resist a poke at earlier explanations implicating family dynamics: "Freud thought that a distant, emotionally cold father might prevent a boy from identifying with Dad and steer him to homosexuality," they wrote. "How much stranger it will be if, instead of the father's psychological rejection, it is the mother's immunological rejection that inadvertently but actively makes her son gay?" Strange indeed, and potentially significant, given a public discussion that can still become charged over the question of whether homosexuality is a "choice" or an inborn characteristic. If Bogaert and Blanchard's hypothesis is correct, for at least some gay men it is not either, exactly, but the result of an experience they encountered before birth.

A third question about the origins of adult characteristics is posed by the Pima Indians of the Gila River Reservation in Arizona. The tribe can claim an impressive if undesirable distinction: they have the highest incidence of type 2 diabetes in the world. More than half of its population over thirty-five suffers from the disease, two and a half times its prevalence in the U.S. population. The Pima's susceptibility to diabetes is often ascribed to genes.

Their ancestors evolved as hunter-gatherers, goes the familiar explanation, and their inherited physiology is ill-suited to life on the reservation. There is little doubt that the high incidence of diabetes among the Pima, and among Native Americans in general, has a significant genetic component. But recent research has added another potential contributing factor: prenatal experience.

During pregnancy, a diabetic woman's high blood sugar may disrupt the developing metabolism of the fetus, predisposing it in turn to diabetes and obesity. Research has firmly established that the offspring of diabetic women are themselves more likely to have diabetes—in one recent study, *seven times* more likely. But how can we know whether this intergenerational transmission of diabetes risk stems from DNA inherited at conception, or from conditions experienced in the womb? The answer may be found in a study that has followed a large group of Pima Indians since 1965, a long-term effort to understand how diabetes is passed down in the tribe and how this vicious cycle may be stopped. Dana Dabelea is an associate professor of epidemiology at the University of Colorado, Denver, and one of the study's investigators. "Our data indicate that prenatal exposure to the mother's diabetes confers a risk of diabetes in the offspring that is over and above any genetic susceptibility," she tells me. "In fact, we find that exposure to a diabetic intrauterine environment is responsible for about 40 percent of the diabetes in the children in our study." Exposure to maternal diabetes in utero accounts for most of the increase in type 2 diabetes among Pima children over the last thirty years, she says, and may well be a factor in the alarming rise of the disease nationally. Dabelea does find a ray of hope in her studies of the Pima: "If we could intensively control diabetic women's blood sugar during pregnancy, we could really bring down the number of children who go on to develop diabetes."

The multigenerational scourge of diabetes among Native Americans has also drawn the attention of anthropologists like Daniel Benyshek, a professor at the University of Nevada, Las Vegas. "I'm interested in how people think about their diabetes, how they understand the disease in the context of their lives," he tells me. In his studies of the Pima and other Native American tribes, Benyshek has found that those who believe that diabetes is their genetic destiny tend to hold fatalistic attitudes about the illness: "They say things to me like, 'I'm going to get it anyway, so what's the point of exercising or watching my diet?' or 'I know I'm going to die of this, so I'm going to eat and drink whatever I want.'" David Kozak, an anthropologist who has worked with the Pima at Gila River, has named this set of attitudes "surrender." Almost 80 percent of the Pima interviewed by Kozak identified genes as the cause of their community's diabetes epidemic; many take this to mean that the disease is "in Indian blood," and so is their unavoidable fate.

When Benyshek has shared research findings about the potential fetal origins of diabetes with tribe members, he has noticed a quite different reaction. "The idea that some simple changes made during pregnancy could reduce the offspring's risk for diabetes fosters a much more hopeful and engaged response," he says. Benyshek believes that programs aimed at improving the diet and exercise habits of pregnant women could be an effective complement to existing interventions, which may paradoxically increase participants' sense of helplessness by emphasizing Indians' genetic vulnerability. "Young women, in particular, are enthusiastic about the idea of intervening in pregnancy to break the cycle of diabetes," he says. "They say, 'I tried dieting, I tried exercising, and I couldn't keep it up. But I could do it for nine months, if it meant that my baby would have a better chance at a healthy life.'"

* * *

In the weeks following my revelation on a SoHo street, I keep seeing the people around me in a disorienting double perspective, as if I'm wearing the kind of X-ray glasses that used to be advertised in the backs of comic books: Amazing and Surprising Fetal Spectacles! On this day in mid-January, I'm on my way to yet another obstetrician checkup. It's too cold to walk, so I'm taking the bus across town; as we jounce along, I study the faces and bodies of the people sitting nearby. The heavyset woman leaning against the window, the rail-thin teenager fidgeting with his iPod—were their differing metabolisms programmed by their prenatal experience? The frail-looking man perched painfully on one of the seats for the disabled—could his illness have originated in his mother's womb? The notion that any of us could be affected in adulthood by conditions we encountered before birth strains belief.

And yet, the idea that we owe anything about our mature selves to our emotional experiences during childhood was once considered preposterous, too. When Freud first floated the theory that children's early relationships with their mothers and fathers had lifelong consequences, he encountered incredulity, ridicule, even condemnation. Now, childhood is the first place we look for the roots of our resilience or our pathology, whether in therapy or on talk shows or in memoirs of authors' early years. The theory of fetal origins faces a decidedly more daunting bid for popular acceptance, however. For one thing, our belief in the formative power of childhood experiences relies on memories we can recall, explore, process (sometimes *ad nauseam*). The possibility that we may have been shaped by experiences about which we can have no memory, and almost no knowledge, is simply confounding, and therefore easily dismissed.

There's a way, too, in which the theory of fetal origins doesn't conform to our preferred explanations of ourselves. One such favored understanding concerns our lifestyles as adults: if we only ate better and exercised more, we believe, we wouldn't have heart trouble, or diabetes, or those extra pounds around our hips. We tolerate, even seek out, the exhortations of fitness gurus and the harangues of weight loss experts because they imply the possibility of change—if not right this minute, then next week, or next New Year's. Paradoxically, we also relish genetic explanations: they seem to hand us our biological marching orders, relieving us of the responsibility of command. To hear that there's a gene "for" character flaws like impulsivity or neuroticism is oddly soothing, a balm to overburdened consciences: there's nothing more to be done. Prenatal influences, on the other hand, raise only an unsettling kind of contingency: things might have been different, but they weren't, and now it's too late. Fetal life was, for most of us, a dispiritingly long time ago, a distance that only adds to the implausibility of the exercise.

This, in the end, may be the theory's highest hurdle to acceptance: its sheer improbability. We react to the notion of fetal origins with something like Teddy's proto-adolescent eye roll: *Yeah, right, whatever.* But maybe the idea just hasn't been around long enough. With time, it could come to seem eminently plausible, even inevitable; it could become part of our mental wallpaper. Before that can happen, though, we'll need to see some hard evidence.

In a research lab located just outside Boston, there is a freezer filled with the physical artifacts of more than two thousand pregnancies: vials of blood drawn from women during pregnancy;

tubes filled with their babies' umbilical cord blood, collected immediately after delivery; more blood samples from women and their offspring, taken when the children reached three and then seven years of age. This corporeal evidence is supplemented by voluminous records kept on the mother-child pairs, beginning in pregnancy and extending through childhood. The files include women's weight and blood-pressure readings during pregnancy, as well as the results of blood tests and details of labor and delivery; the answers to detailed questionnaires about women's diets, physical activity, and home environment during pregnancy; the scores on developmental tests given to children at periodic intervals beginning in infancy; children's medical diagnoses, medication use, and bodily measurements.

All this information has been gathered for Project Viva, a pioneering study of prenatal influences on later health. This longitudinal cohort study—that is, one that follows a single group of people over many years—is the brainchild of Matthew Gillman, professor of population medicine at Harvard Medical School. I meet with Gillman one afternoon in his sunny office in a brick medical building in Boston. Balding and bushy-browed, Gillman has a steady intensity leavened by an appealingly dry sense of humor. In the mid-1990s, he became intrigued by the emerging idea that what happens very early in life can have effects on the health of babies, children, and perhaps even adults. "When I first read the work of David Barker on birth weight and later health, I was skeptical. There's a lot that happens between birth and adulthood, and I wasn't convinced that prenatal influences had much impact," Gillman tells me. But he kept reading, and within a few years he published an article charting the evolution of his thinking, titled "The Fetal Origin of Adult Disease: From Skeptic to Convert." Gillman wanted to conduct a study of the effects of childhood experience on later health, "but Barker had started me wondering: when does childhood really begin? I think it begins before birth, and so my study would have to start there, too."

With a grant from the National Institutes of Health, Gillman began recruiting subjects for Project Viva in 1999, ultimately enrolling 2,670 pregnant women. All of the participants in the study were members of a Harvard-affiliated health-maintenance organization, and so (with participants' permission) Gillman and his staff were able to have unusual access to their subjects' complete medical records in addition to the questionnaires and biological samples gathered specifically for the study. From the beginning of Project Viva, Gillman decided to focus on the early-life origins of three outcomes in particular: asthma and allergies, neurocognitive development, and obesity and heart disease. The study has already yielded a trove of intriguing findings.

Such as: The children of women who have a higher intake of vitamin D during pregnancy are less likely to show early signs of asthma. A study of 1,194 Project Viva mothers and their three-year-old offspring, published in 2007 in the *American Journal of Clinical Nutrition*, found that women who consumed the greatest amount of vitamin D had children with the lowest risk of asthma. This was the case whether the nutrient came from supplements or from dietary sources like liver, eggs, and dairy products.

And: Eating a lot of fish during pregnancy seems to produce smarter kids. A study of 135 Project Viva mothers and their six-month-old babies, published in 2005 in the journal *Environmental Health Perspectives*, found that greater fish consumption during pregnancy was associated with better infant cognition. The highest scores on a test of visual recognition memory were found among the offspring of women who ate more than two servings of fish a week during pregnancy, but had relatively low levels of mercury in their blood; similar results were found when another group of 341 Project Viva children were tested at three years of age. Pregnant women should eat plenty of fish, concluded Harvard Medical School assistant professor and Project Viva investigator Emily Oken, but they should choose varieties low in mercury and high in the "good" omega-3 fats, such as sardines and salmon.

And: Women who gain less than the recommended amount of weight during pregnancy are less likely to have overweight children. A study of 1,044 Project Viva mothers and their children found that greater weight gain during pregnancy was associated with a higher body mass index in three-year-old children. Women who gained excessive or even appropriate amounts of weight, according to the guidelines set by the Institute of Medicine, were four times more likely to have an overweight toddler than were women who gained less than the IOM advises (between twenty-five and thirty-five pounds for normal-weight women). "New recommendations for gestational weight gain may be required in this era of epidemic obesity," concluded the study—and indeed, data from Project Viva were instrumental in devising the most recent set of recommendations, which places a stricter limit on the amount of weight obese women should gain during pregnancy.

The goal of Project Viva is not simply to generate research findings, Gillman says, but to devise effective interventions during pregnancy to improve the later health of the offspring. He knows this is a tall order. "Behavior is not easy to change. And pregnant women may encounter social and behavioral contexts that make change even more difficult—they may live in communities where fast-food restaurants far outnumber stores selling fresh fruits and vegetables, for example." Still, he says, "Women want to do the best they can by their fetuses, and pregnancy is a time when they may be more open to change. If we can intervene in the right way, we can achieve beneficial health outcomes without freaking women out." On this point, he is adamant: "This is not about blaming women or making them more anxious than they already are."

Gillman's colleague Emily Oken agrees. "There does seem to be a lot that a woman can do during pregnancy to improve the health of her offspring, but that doesn't mean we should place all the responsibility on her shoulders," she tells me. "For one thing, the mother's physiology and behavior may well have been shaped by *her* prenatal experience, and the same with her mother, so how far back do you go?" And, Oken points out, prenatal experience isn't the end of the story. "If a woman gains a lot of weight during pregnancy, for example, that doesn't mean that her offspring are doomed to a lifetime of obesity. It means that they've been pointed on a particular trajectory, but if they go on to eat sensibly and exercise—perhaps with the help of an intervention program—that trajectory can be altered. So, our research can help target for intervention children who are at higher risk because of their fetal experience."

Right now, Project Viva is the only longitudinal cohort study of pregnancy and birth outcomes in the United States. But a much bigger investigation, "the eight hundred pound gorilla," in Gillman's words, is just getting underway. This is the National Children's Study, which will enroll one hundred thousand pregnant women from all over the United States, following their offspring from before birth to age twenty-one. Researchers from the National Institutes of Health will conduct interviews with the women about their habits and behaviors during pregnancy, sample their hair, blood, saliva, and urine, and test the water and dust in their homes. The first results from the study, concerning the causes of premature births and birth defects, are anticipated in 2012.

Project Viva and the National Children's Study are the kind of investigations that are beginning to bring us answers about the impact of fetal experience on later life. Other such studies will likely be initiated as interest in the theory of fetal origins grows. But conducting this sort of research poses unique challenges, as Matthew Gillman and his colleagues have discovered. In fact, their experience with Project Viva provides an early look at the obstacles facing an emerging generation of fetal researchers. Their first and most basic objective will be simply hanging on to their studies' subjects. During pregnancy and early child rearing, families may move to bigger houses and relocate for better jobs; the daily scramble to get to work and school may push aside the regular questionnaires and child visits that such studies require.

The staff of Project Viva has worked to make participation as easy and convenient as possible, offering weekend hours, performing home visits, and conducting a marathon series of child visits during school holidays. They have learned to collect extensive contact information from each subject: not just street addresses, email addresses, and home and cell phone numbers, but the names and numbers of three friends who can be consulted about the participants' whereabouts. (If all else fails, the Project Viva staff will Google a study subject who's fallen out of touch.) And they strive to create a friendly sense of community among the participants, sending out periodic newsletters apprising them of the latest findings and noting fun facts like the most popular names for "Viva babies" (Julia for girls, Andrew for boys). When I visited Gillman, he and his staff were planning a celebration of the study's tenth anniversary, to which all the subjects would be invited.

Another challenge is retaining the study's focus and purpose even as it is allowed to evolve and grow in unanticipated directions. The shape of the project shifts as investigators incorporate new findings into their next round of questions: science as rolling improvisation. "You always wish, belatedly, that you'd asked this question or requested that sample, but with a longitudinal cohort study you just have to move forward with what you have," says Gillman. And, of course, the subjects themselves are constantly changing. The oldest of the Viva kids are now ten years old, and Gillman is already looking ahead to their burgeoning adolescence. "With the teenage years will come new health issues, new behaviors—sex, drugs, and rock 'n' roll," he says. Gillman and his staff are considering new ways to engage teenage participants: asking them questions directly, instead of querying their mothers, or perhaps starting a Project Viva Facebook page.

He plans to keep the study going as long as possible. Some of the most interesting results, he predicts, will come only as subjects age. This brings up the delicate matter of the project's continuing leadership. Gillman, who is in his mid-fifties, addresses the unspoken question head-on: "The definition of a successful cohort study is that it outlasts the original investigator," he likes to say. He feels confident that Project Viva has already overcome at least one hurdle: persuading the scientific community, and funding organizations, that research on fetal origins is worthwhile. "The notion that prenatal conditions have an effect on later health is gaining wider acceptance," he notes. "Ten years ago, this was a new and quite controversial idea. Now, it's got a foothold. People are starting to believe that what happens in the womb can affect us for a long time after."

For now, such intensive scrutiny of fetal life is limited to the small number of individuals enrolled in scientific studies. But it's possible to imagine a day when all fetal lives are exhaustively documented: thousands of freezers stocked with samples of fetal blood, archives filled with ultrasound photographs and videos, files fat with detailed records of what women did and ate and felt during pregnancy. Already, the physical artifacts of fetal life—the pottery shards and flint arrowheads of a mysterious world—are yielding specific clues about what went on during the nine months of that world's existence.

As we've seen, umbilical cord blood collected immediately after delivery is being analyzed for the presence of environmental toxins. When such substances are found—and one recent study discovered an average of two hundred industrial chemicals and pollutants in the cord blood of newborns tested—it can give us a picture of the exposure of women and fetuses during pregnancy. Likewise, the placenta, once discarded immediately after birth, is increasingly being preserved and analyzed for insight into the environment that prevailed inside the womb. The size and condition of the placenta can reveal how well it functioned during pregnancy, and provide information about the cause of conditions like intrauterine growth restriction and cerebral palsy. The College of American Pathologists now recommends that hospitals store placentas from all births for at least three days, and immediately test those that appear abnormal or that come from a high-risk pregnancy or birth. Placentas have even become evidence in the courtroom. Doctors sued for malpractice have introduced the organ at trial to show that a placental defect, and not physician error, was the cause of a baby's injury or illness.

Amniotic fluid, too, is being collected and scrutinized for the story it has to tell. Simon Baron-Cohen, a professor of developmental psychopathology at Cambridge University in England, is well known for his research on autism, including his controversial theory that the condition is a kind of hypermasculinity of the brain. In an effort to trace the roots of behaviors associated with autism, Baron-Cohen initiated the Cambridge Fetal Testosterone Project. He and his team have tested the amniotic fluid of about two hundred fetuses to determine levels of the male hormone testosterone (testosterone is present in the amniotic fluid of fetuses of both sexes, though female fetuses produce much less). In the years since these babies were born, the investigators have evaluated the children at periodic intervals, looking for links between fetal testosterone levels and later child characteristics. "Amniotic fluid provides a window into the child's past—a chemical record of that child's time in the womb—that allows informed prediction about the child's future brain, mind, and behavior," Baron-Cohen tells me. So far, he has found that high levels of fetal testosterone are associated with reduced eye contact at one year of age, more limited vocabulary at two years of age, more social difficulties at four years of age, and greater difficulties with empathy at eight years of age. Higher levels of fetal testosterone were also associated with more narrow interests in childhood, a stronger interest in systems (such as finding out how things work), and a greater number of autistic traits. "This is not the retrospective speculation of psychoanalysis," says Baron-Cohen, "but solid biological evidence of the conditions a child experienced at the very beginning of his life." Baron-Cohen and his team are now embarking on a much larger study, in collaboration with the Danish Biobank, to test if fetal testosterone levels are elevated in children who are later diagnosed with autism or Asperger's syndrome.

As we come to know more about what actually happened during the nine months of a particular fetus's gestation, what will we do with this information? As Emily Oken of Project Viva suggests, we may use it to identify candidates for intervention: the children of women who had diabetes or gained excessive weight during pregnancy, for example, could be offered guidance on exercise and weight management. It's possible, too, that one day the medical care of children and adults will be tailored to the conditions they experienced before birth. Oken puts me in touch with Mary-Elizabeth Patti, an assistant professor at Harvard Medical School and a physician-scientist at Harvard's Joslin Diabetes Center. "I always ask my patients what their birth weight was," Patti tells me. "Patients are often surprised at the question—they're expecting me to ask about their current lifestyle. But we know that low-birth-weight babies become adults with a higher risk of diabetes, so having that information gives me a more complete picture of their case." Patti is now researching how information about patients' birth weights could translate into tailored courses of treatment.

Farther off in the future, such behavioral approaches may be joined by pharmaceutical ones. Working with animal models, some scientists are experimenting with ways to chemically reverse the consequences of adverse prenatal conditions. By providing pregnant rats with an antioxidant, for example, researchers at the University of Montreal prevented the development of high blood pressure in their adult offspring. By giving pregnant rabbits compounds that block a particular enzyme, scientists at Northwestern University protected the rabbit fetuses from brain damage caused by lack of oxygen. By injecting the hormone leptin into newborn rat pups, investigators from the University of Southampton in England and the University of Auckland in New Zealand stopped the rat offspring from becoming obese. By feeding pregnant rats genistein, a compound found in soy, researchers at Duke University offset the negative effects of exposure to bisphenol A in their offspring. And by administering the nutrient choline to rat pups prenatally exposed to alcohol, scientists at San Diego State University reduced the severity of the young rats' learning deficits. (Zinc supplements given to pregnant mice have also been shown to have a protective effect on mice fetuses exposed to alcohol.) The application of such treatments to humans is a long way off, but the premise is a promising one: following a less than ideal fetal life, it's possible to get a do-over.

Perhaps our burgeoning knowledge about fetal life will even shape how we think about our mature mental and emotional characteristics: why this individual is so intelligent, why that person is so prone to anxiety, whether this adult is fully responsible for his actions. If that sounds far-fetched, consider the case of Charles Gaston. After he was convicted in 1989 of killing a clerk at a cigar store in Sacramento, California, Gaston faced the prospect of the death penalty. At his sentencing hearing, his lawyer presented evidence that he had been exposed to alcohol in utero; according to his adoptive mother, the social worker who handled the placement told her that Gaston's biological mother was drunk even on the night she gave birth to him. After listening to medical experts' testimony about the effects of fetal alcohol syndrome, the judge granted Gaston a measure of clemency, sentencing him to life in prison rather than execution.

To be sure, many others accused of crimes, including the notorious double murderer Robert Alton Harris, have pleaded unsuccessfully for leniency based on FAS. (Then-governor of California Pete Wilson rejected Harris's request for a stay of execution even as he called heavy drinking by pregnant women "nothing less than child abuse through the umbilical cord.") Whatever they decide, the question judges are weighing in these cases is startling to contemplate: nothing less than whether a person's prenatal experience can affect his culpability for crimes he committed as an adult.

If some worry that the consequences of prenatal experience will be invoked too freely (legal scholar Alan Dershowitz has called fetal alcohol syndrome an "abuse excuse"), others may well be concerned that they will be understood too rigidly. The eminent Harvard psychologist Jerome Kagan has written of the dangers of "infant determinism": the misguided assumption that the experiences of early childhood forge a permanent template for the rest of life. Could a new generation of scientific findings lead us to what we might call "fetal determinism"? One who is apprehensive about this possibility is Darshak Sanghavi, a physician and author who writes about medicine for the *New York Times* and the *Boston Globe*. On the morning after my conversation with Project Viva investigators, I meet Sanghavi for breakfast at a deli in the Boston suburb of Brookline. Sanghavi, friendly and alertly inquisitive, is a pediatric cardiologist at the University of Massachusetts Medical School; he says he already sees the notion of fetal origins creeping into his practice, and it troubles him. In fact, I first became aware of Sanghavi's views when he critiqued a magazine article I'd written on the subject.

"Turning to the womb to explain complex social and public-health problems ultimately means people have given up on changing the things that really matter. That's too bad," Sanghavi wrote in a commentary in the online magazine *Slate*. "The truth is that nothing in this world worth having comes easy. And as any hard-working student who made it to college, overweight person who's changed his or her lifestyle, or adult who's worked through depression can tell you, at some point you have to stop blaming your issues on your mother's uterus." As he elaborates to me over eggs and oatmeal, he's concerned that if we become convinced that our destinies are determined in the womb, society will no longer be willing to invest in the welfare of individuals after they're born. "Why bother funding children's health initiatives or universal preschool, if physical and cognitive functioning have been set in utero?" he asks.

Sanghavi's warning is a useful one. Prenatal experience doesn't force the individual down a particular path; at most, it points us in a general direction, and we can take another route if we choose. Imagine water flowing downstream: prenatal influences might dig a canal, so to speak, making it easier for the water to flow one way rather than another. But with the effort Sanghavi describes, we may be able channel our fates in a different direction. The theory of fetal origins ought to contribute to complexity, not reduce it; if we take care in how we think about prenatal influences, they may add another layer to our understanding of who we are and how we got to be this way.

Once home from Boston, the third-trimester urge to nest takes hold with a vengeance. Walking around Manhattan, scanning the shop windows, my eye is caught not by a dramatic dress or a pair of sharp-toed shoes but by a wooden pull train or a corduroy jumper. Children's stores exert a magnetic attraction, and after being pulled into yet another I emerge with a baby book: a twin to the one I bought for Teddy, now home to his birth announcement, his hospital bracelet, an inky print of the soles of his feet. At home, I paste an image on the first page of the new book: it's a black-and-white shot, in profile, of my second son's fetal self. He and his peers will know far more than any previous generation about what their lives were like before birth. But what about those of us for whom fetal life was long ago? David Barker likes to say that "everyone has been changed by their experience in fetal life"; I want to know how. So I go to the expert, the indisputable authority on the subject: my mom.

On a Sunday afternoon near the end of January, I'm sitting in the kitchen of the house in suburban Philadelphia where I grew up. My mother is making dinner, and I'm asking her about our erstwhile joint venture: her pregnancy with me. On the kitchen table I've spread out the meager physical evidence of my fetal life. There's a picture of my mother pregnant, her belly round and inscrutable below her open, smiling face. There is the obstetrician's record of my birth, noting a normal delivery and a solidly average birth weight: seven pounds, seven ounces. And there is my own first picture, taken not in the womb but on the hospital's flowered sheets. In the photograph I look seriously disgruntled, a mood I'm revisiting now as my questions encounter one unsatisfying answer after another.

"What did you eat when you were pregnant with me?" I ask my mother, who's standing at the stove, stirring a pot of tomato sauce.

"Oh, you know, healthy things," she says. "With you, I craved grilled cheese sandwiches, and with your sister, I wanted spaghetti with garlic and oil." She adds a dash of salt to her simmering sauce.

"Did you exercise while you were pregnant?"

She snorts, and tosses in a handful of oregano. "Only running nuts exercised back then."

"Did you feel happy?"

"Oh, yes," she says, vaguely. She dips her spoon in for a taste, wrinkles her nose, adds more salt.

"Well, what else can you tell me about your pregnancy with me?" I demand, what I recognize as a rising note of adolescent petulance in my voice. Another reason we resist the theory of fetal origins now occurs to me: it reminds us of our onetime physical union with our mothers, an almost unfathomable intimacy that can be deeply uncomfortable to contemplate. As grown adults, the idea that we were once so profoundly close, so completely dependent, on another individual is not always welcome; nor is the notion that what that person did decades ago could still affect us now.

My mother dips a spoon into the sauce, tastes it, holds it out to me: "Want to try?"

I heave a sigh. I'm quickly starting to feel ridiculous about this whole exercise, as if I'm Shirley MacLaine channeling past lives as Charlemagne's lover or as an inhabitant of the island of Atlantis. It's time to acknowledge the disappointing reality: for me and for everyone who is now an adult, fetal life is indeed a lost continent—unrecoverable, sunk in the watery deep. It will always be an absence, an indisputably important period about which we can speculate but never know.

This is largely true even for those being born today—people like my second son, surveilled in utero since he was a mere embryo, a lima bean with a beating heart. We know much more than we once did about fetal life, both in general and in the particular case, but still remarkably little. When I speak to researchers, even those who've been working in this field for decades, I'm struck by how mysterious the fetus remains to them: it's a creature more rumored than real, encountered only in shadowy glimpses and amplified rumbles. Our body of knowledge about the fetus enlarges with every experiment and observation these scientists publish. Even so, our ability to control the course of fetal life will never be close to complete. It's a difficult lesson of parenthood that we must now apply to pregnancy: all we can do is try our best, and wait to see how it turns out.

EIGHT MONTHS

Waiting in line at the drugstore, shifting from foot to foot to find a comfortable position, I have the eerie sensation that I'm looking in a mirror. The magazine rack beside the checkout counter is filled with pictures of pregnant bodies. Not just the parenting publications, but every celebrity and lifestyle glossy seems to feature a bulging belly like mine (well, tauter and better-dressed than mine, but participating in the same biological condition). I know from my own surreptitious perusals that these magazines cover actresses' and models' pregnancies in obsessive detail, from the first "bump" sighting to the second-trimester weight gain to the baby-gear shopping spree. "In this world, pregnancy is considered news, and it is a nine-month news cycle," the editor-in-chief of *Star* magazine has said. Issues featuring pregnant celebrities are often among the magazines' best sellers. The mother of all celebrity pregnancies, of course, was that of actress Demi Moore, who appeared naked and seven months pregnant on the cover of *Vanity Fair* in 1991. Back then, the image still had the power to shock: on its first day on the stands, the issue sold out during the morning rush hour at Grand Central Terminal. Stores in other parts of the country covered Moore's bared belly with a paper wrapper, and several supermarket chains refused to sell the magazine at all. Now, it seems, images of barely clad pregnant women are everywhere we look.

These publications *are* a kind of mirror, in fact, a reflection of a more general preoccupation with pregnancy. Being an ordinary pregnant woman these days is a little like being a celebrity: you're gawked at, commented on, given unsolicited advice— even in a strenuously anonymous city like New York, where the first rule of public behavior is usually *Keep it to yourself.* Pregnant, your anonymity is relinquished; you can no longer pass unnoticed in a crowd. Women used to fending off the gaze of passing men will find that other women, too, now cast an appraising eye over their bodies, and that children stare at them with undisguised interest. Once private, a matter of hushed voices and dim rooms, pregnancy has been brought out into the bright light of public attention and opinion, in books and blogs and television shows. William Liley, the eminent physiologist who performed the world's first fetal blood transfusion, once mused that doctors who chose obstetrics as their specialty were perhaps "compensating for an ungratified curiosity to know where babies come from." It can seem like the rest of us are busy gratifying that primal nosiness, too: we're all obsessed with pregnancy now.

Walking around the city, I keep mental note of the reactions my condition evokes in the faces of passersby: smiles, scowls, gapes, and one fervent "God bless you." Clinging to a pole on a swaying subway car, I feel both gratified and self-conscious about the attention I attract—but mostly, I'd just like a seat. I can't help feeling that we're interested in pregnancy for all the wrong reasons, can't help distrusting the impulse to exhibit and ogle pregnant bodies, using them to sell magazines (as well as a vast new array of pregnancy-related products). I wonder: Will our avid interest in pregnancy lead to greater social investment in the well-being of pregnant women and their fetuses—or will pregnancy remain an occasion to catch a glimpse or make a buck?

It's an important question to ask now, because there is new evidence—from surprising sources—that society's stake in healthy pregnancies is far larger than we knew. Some of the most compelling support for this notion comes, for example, from the story of a virus, an illness that first appeared among grown men.

In September 1918, a deadly strain of the flu began striking enlisted men at Camp Devens, a military outpost in Massachusetts. The first victim showed up at the base hospital on September 8; ten days later, almost seven thousand soldiers had been stricken, their skin turning blue and their coughs bringing up blood. Many of the afflicted died. A colonel who came to inspect the camp reported grimly that "in the morning, the dead bodies are stacked about the morgue like cordwood." Soon, the virus leapt the boundaries of the camp to infect the civilian population. More than a hundred residents of Boston were dead from the flu by the end of the month. The spread of the illness picked up speed, traveling down the coast and across the country. In Philadelphia, officials declared all schools, churches, theaters, and saloons closed in a desperate attempt to contain the virus; still, more than ten thousand people died within a month of influenza's arrival in the city.

After four devastating months, the worst of the pandemic was over, and the work of cleaning up and counting the dead began. But that was far from the end of the story. In addition to the 550,000 Americans who perished in the flu pandemic of 1918, many more—twenty-five *million* more—caught the same vicious virus and survived. Some of the highest infection rates occurred among women of childbearing age, one-third of whom contracted influenza. Douglas Almond, an enterprising young economist at Columbia University, wondered about the fate of the fetuses who were in utero during those dark months of 1918. What became of the pandemic's children?

Almond's office is just a few blocks from my apartment, a lucky thing given my eight-months-pregnant condition. I've now reached a stage that John calls "extreme pregnancy": after a shower, the towel no longer reaches all the way around my middle; my belly precedes me around corners, like a shadow or a dog on a long leash. Today I'm wearing the only thing that still fits me, a sweater that suggests nothing so much as Christo draping a mountain range in black chenille. As I make my way up Amsterdam Avenue on a chilly February morning, I step carefully around icy patches on the sidewalk, feeling enormous and delicate at the same time, like a circus elephant in a tutu. Aware that passersby are watching my fancy footwork, I entertain a fleeting, and probably unfair, thought: If I slipped and fell, would they stop to help me up?

No need to find out; I make it safely to Almond's office, on a high floor of a university tower overlooking Harlem. Almond—boyish, redheaded—meets me at the door, apologizing: he has three children under five, he says, and they kept him awake for most of the night. But he is focused and precise as he launches into an explanation of why he chose to examine the effects of fetal exposure to influenza. "The 1918 flu pandemic offers an exceptional opportunity to evaluate the effects of the prenatal environment," Almond tells me. "It came on suddenly in September of

1918, and had largely dissipated by January of 1919, so groups of individuals born only months apart had very different conditions in utero. This presents what we call a 'severe' test of the fetal origins hypothesis, since it allows us to generate sharp predictions for differences in how these individuals do as adults."

Initially, Almond doubted that the intrauterine conditions provided by a pregnant woman, even one sick with a virulent strain of the flu, could exert any lasting influence on her offspring. "When I started looking at the influenza pandemic, I was skeptical of the fetal origins hypothesis. I didn't think I'd find any long-term effects," Almond says. "But the evidence was the opposite of what I expected." Through an analysis of census data, Almond discovered that those individuals gestated during the pandemic did poorly as children and adults compared to cohorts born shortly before or after the flu hit. "People who were in utero during the pandemic did worse, on average, on just about every socioeconomic outcome recorded," he says. Over their lifetimes, they displayed lower educational attainment, lower income, and lower socioeconomic status; they suffered higher rates of disability, and required higher welfare payments. Individuals gestated during the pandemic were 15 percent less likely to graduate from high school, and 15 percent more likely to be poor; the men earned wages that were 5 to 9 percent lower, and they were 20 percent more likely to have heart disease or to be disabled as older adults. Even their height was affected: when the cohort of people born soon after the pandemic showed up for enlistment in World War II, they were shorter than recruits born the year before and the year after.

It might seem odd for Almond, an associate professor of economics, to be studying the effects of prenatal influences, but he says that fetal origins research is a natural match for the skills and techniques employed in economic analyses. "Estimating the costs of adverse early experience is the kind of thing economists are good at," Almond tells me. "We're used to working with large data sets, and identifying patterns that emerge over time, often very long periods of time." Following his study of the influenza pandemic, he began looking for other real-world situations that would allow him to quantify the lasting effects of prenatal conditions. He found no shortage of such natural experiments.

In an article published in the *Quarterly Journal of Economics,* for example, Almond compared schoolchildren with prenatal exposure to fallout from the 1986 Chernobyl nuclear disaster to those born just before the explosion; his results showed that children in the exposed group were a third more likely to fail middle school. The young people Almond studied lived in Sweden, more than a thousand miles from the Chernobyl reactor in the Ukraine. His findings suggest that prenatal exposure to nuclear fallout, even from very far away, can impair offspring's cognitive ability "at radiation doses currently considered harmless," he wrote.

In a working paper issued by the National Bureau of Economic Research, Almond compared individuals who were gestated during the period of scarcity that accompanied China's Great Leap Forward to those who were in utero just before or after the famine. Members of the first group, Almond found, have lower literacy levels and labor market status, live in smaller houses, and make less advantageous marriages as adults.

Almond also evaluated the effects on fetuses of fasting during the Islamic holy month of Ramadan, comparing the infants of Arab-American women whose pregnancies overlapped with the Ramadan fast to babies in the same Michigan community who were gestated during other times of the year. His results, also reported in an NBER working paper, demonstrated that women who fasted while pregnant gave birth to babies who were born earlier and had lower birth weights. A similar study by Almond of Muslims in Uganda found that individuals who were born nine months after Ramadan were 22 percent more likely to be disabled as adults, with higher rates of vision, hearing, and learning problems.

Almond is not the only economist using the powerful, precise tools of his discipline to gauge the effects of prenatal experience. Increasingly, economists are applying their massive data sets, sophisticated statistical techniques, and ingenious experimental designs to phenomena like premature birth, low birth weight, and fetal exposure to toxins, toting up their cost over lifetimes and across populations. Their analyses reach far back in time and deep into data: birth registers, census figures, military records, death certificates. They return with provocative evidence that when pregnancies go well, the benefits are shared by us all—and when prenatal conditions are poor, the price paid by society can be high.

That's certainly the case for the youngest victims of the flu pandemic of 1918. The cost of their prenatal exposure to the virus was significant not only for the individuals themselves, but for society at large: their intrauterine experience rendered them less able citizens and less productive workers. These "children" of the pandemic are now in their nineties, but the effects of their time in the womb are still unfolding. Because of the long-lasting impact of prenatal conditions, Almond tells me, "you could say that the influenza pandemic of 1918 isn't over yet."

Walking back down Amsterdam, the icy patches thawing now in the wan February sun, I think about how fetal origins warps our sense of time, much as pregnancy itself does. To conceive a child in the hazy shimmer of summer, and give birth to him when the ground is slick with ice, makes my notion of time soften and bend. Fetal origins, too, stretches time like one of Salvador Dali's melting clocks, moving cause and effect so far apart that the whole landscape starts to look new and strange. As with Almond's flu pandemic study, the work of his economist peers on the effects of early-life experience is producing novel explanations for familiar facts, fresh insights that contribute to a growing awareness of the importance of pregnancy.

It's a well-established finding, for example, that taller people on average make more money. The writer Malcolm Gladwell examines this phenomenon in his 2005 book *Blink*, concluding that it's a product of what he calls "the Warren Harding error": a biased judgment made so quickly we are not even aware of it. It was just such a baseless prejudice in favor of tall people, Gladwell writes, that led American voters to elect to the presidency Warren Harding, a tall but otherwise undistinguished individual. According to Gladwell's own survey, male CEOs of Fortune 500 companies are three inches taller than the average American man; he cites research demonstrating that each additional inch of height above the average is worth $789 per year in salary. All this shows, says Gladwell, that we are "absurdly biased in favor of the tall." A universally shared "unconscious bias" gives tall people an undeserved advantage in hiring and promotion, he writes: "We see a tall person and we swoon."

Economist Anne Case has a different explanation for the greater success of tall people. "Height is positively associated with cognitive ability, which is rewarded in the labor market," Case tells me when I reach her at her office at Princeton University. Tall people earn more, that is, because they are smarter— and both height and intelligence, she notes, can be affected by conditions in the womb. Case and a Princeton colleague, economist Christina Paxson, examined the height records and test scores of two large groups of Britons and Americans who had been tracked by researchers from birth to adulthood. Case and Paxson's results, published in the *Journal of Political Economy*, show that on average, taller children score higher on tests of cognitive ability, and that these superior cognitive skills explain a large portion of the larger wages earned by tall adults.

Height is mostly genetically determined, of course. But adult height is also sensitive to environmental conditions experienced early in life. The quality of the intrauterine environment—including the pregnant woman's health and nutrition, her use of cigarettes, alcohol, and other drugs, and the fetus's exposure to infections and toxins—influences the height attained by offspring (as do nutrition levels and disease exposures later on in childhood). These same factors also affect cognitive ability. "It may be that employers aren't wrong to generally favor taller workers, at least until the employees have a chance to demonstrate their abilities," Case tells me. "Our study suggests that height and cognitive ability are linked, most likely because they are both influenced by the same set of factors operating before birth."

It's clear that the conditions we experience both before and after birth work together to affect our intelligence, notes Howard Gardner, professor of cognition and education at Harvard University and author of the influential theory of multiple intelligences. "Compare a child who has a dozen healthy experiences each day in utero and after birth to another child who has a daily diet of a dozen injurious episodes," Gardner writes. "The cumulative advantage of a healthy prenatal environment and a stimulating postnatal environment is enormous." But just how, and how much, intelligence is affected by prenatal conditions is hotly contested, as is almost every assertion in the field of intelligence research. In a 1997 paper published in a major scientific journal, *Nature*, Bernard Devlin and two coauthors made a landmark attempt to determine the contribution of the fetal environment to IQ. Devlin, an associate professor of psychiatry at the University of Pittsburgh School of Medicine, and two colleagues analyzed 212 studies of IQ, including many studies of twins, the type of experiments often used to estimate how much impact genes have on intelligence. Charles Murray and Richard J. Herrnstein, authors of the controversial book *The Bell Curve*, depended on such studies to reach their conclusion that the heritability of IQ is 60 percent (that is, 60 percent due to genes, and 40 percent due to childrearing environment). But that approach, Devlin pointed out, did not factor in twins' *other* environment: the one they shared in the womb. The model that fits the data best, he reported, takes prenatal conditions into account. When these are included, the intrauterine environment accounts for 20 percent of IQ similarity between twins, and genes for only 34 percent.

The impact of the prenatal environment on intelligence is equal to if not greater than the impact of a child's upbringing, Devlin contended at the time of his paper's publication, adding that improved prenatal care and nutrition for disadvantaged pregnant women might raise their offspring's IQ. (One of Devlin's coauthors was Kathryn Roeder, a Carnegie Mellon University statistics professor who is also Devlin's wife. She was pregnant at the time they were writing up the study's conclusion. "It made me even more careful" about taking care of her pregnancy, she told a local newspaper reporter, joking, "Each time I ate my brown rice, I thought, 'There's another two IQ points.'")

Not all intelligence researchers share Devlin's views, but a more recent twin study provides support for the notion that the prenatal environment has an influence on IQ. In a 2007 paper published in the *Quarterly Journal of Economics*, UCLA economist Sandra Black analyzed an unusually extensive data set on twins born in Norway between 1967 and 1987. Noting that the difference in twins' birth weights is largely due to nutritional intake in utero, she sought to determine the relationship of birth weight and intelligence. On average, Black found, the twin born at the higher birth weight scored significantly better on IQ tests (and, harking back to the work of economist Anne Case, was significantly taller).

Black's study indicates that the superior prenatal conditions enjoyed by one twin can advantage him over his sibling in ways that last into adulthood. Could the same be true for entire populations? For decades, black Americans have lagged behind whites on a wide variety of measures: infant mortality, physical health, longevity, educational attainment, workplace earnings. Even as opportunities for African Americans have expanded, and more egregious forms of discrimination have faded, these disparities have remained stubbornly persistent. It is also the case that African American pregnant women are more likely to have poor nutrition, to suffer from depression and anxiety, to be exposed to environmental toxins, and to experience trauma. "Fetal health may be at the fulcrum of black-white disparities in life outcomes," says Douglas Almond. If blacks' and other minorities' disadvantage begins before birth, then efforts to reduce racial disparities in adult health and economic outcomes could achieve results by focusing on the prenatal environment.

Almond's own research has demonstrated that programs that improve conditions for pregnant women can have a positive influence on their babies, and that this effect may be especially pronounced for minority women. Seizing on yet another natural experiment, Almond examined the impact of the introduction of the food-stamp program in the late 1960s and early 1970s. The program was rolled out on a state-by-state basis, allowing Almond to compare birth outcomes for poor women who received food assistance during pregnancy to those who did not. His results, published in *The Review of Economics and Statistics*, found that women who were enrolled in the program three months before they gave birth delivered babies with higher birth weights, and that the improvement was especially significant for African Americans. In recent years, Almond notes, early-life health measures of blacks have stagnated; black infants are two and a half times more likely to have low birth weight as white infants, and are more than twice as likely to die before age one. Given the potentially lasting effects of prenatal experience, Almond warns, it may be the case that "a future of racial inequality is being programmed."

The dystopian ring of Almond's words stirs a memory of a long-ago college seminar. From a shelf I pull my dog-eared copy of *Brave New World*, the 1932 novel by English writer Aldous Huxley depicting a nightmare future of haughty "Alphas" lording over groveling "Epsilons." I remember being taught that *Brave New World* was about the dangers of eugenics, of the notion—popular in Huxley's time—that some were destined to rule by virtue of their superior genetic material. Reading it again now, it seems instead like an eerily prescient exploration of the significance, for good or ill, of fetal experience.

When the book opens, we are at the Central London Hatchery and Conditioning Centre. The Director is giving a group of students a tour of the Hatchery, a bustling laboratory where human embryos are prepared for their future roles in society. The lab's scientists don't content themselves with "merely hatching out embryos: any cow could do that," the Director explains. "We also predestine and condition. We decant our babies as socialized human beings, as Alphas or Epsilons." He shows the pupils how bottles containing Alpha embryos are tended with care, readying their occupants to become thinkers and leaders. The Epsilon embryos, by contrast, are deprived of oxygen, dosed with alcohol, and bombarded with X-rays, rendering them fit to do menial jobs like cleaning sewers.

"Nothing like oxygen-shortage for keeping an embryo below par," the Director notes with satisfaction, rubbing his hands together.

"But why do you want to keep the embryo below par?" a student asks.

"Ass!" the Director hisses. "Hasn't it occurred to you that an Epsilon embryo must have an Epsilon environment as well as an Epsilon heredity?"

Closing the cover, I find myself wondering if we are inadvertently running an experiment akin to those carried out in Huxley's hatchery. Are we "predestining" our fetuses for their roles in life, exalted or humble, by the conditions we provide to them before birth? And as we learn more about the effects of prenatal experience, will the arms race of upper-middle-class parenting—ever more resources, more enrichment, more pressure—move into the womb, creating a class of aspiring Alphas advantaged from the start?

Looked at another way, however, fetal origins research could be used to reduce rather than to exaggerate social inequality, through interventions that improve the lot of all pregnant women and their fetuses. It's happened before. The cleaner water, bolstered nutrition, and more advanced medical care of the modern era conferred their most significant benefits on pregnant women and fetuses, improvements responsible in large part for the better health and longer lives we now enjoy as adults. Dora Costa, an economist at UCLA, estimates that enhanced prenatal and early postnatal conditions account for at least 16 percent of the leap in life expectancy that occurred over the course of the twentieth century, from forty-nine years in 1900 to seventy-seven today.

Indeed, long before the emergence of the fetal origins hypothesis, leaders in many countries suspected that strong adults— and, of greatest interest to them, strong armies—begin with healthy mothers. The creation of public programs to improve the health of women and babies began in nineteenth-century Europe, where governments found themselves in need of robust young men to fight their wars and expand their empires. Following a crushing defeat in the 1871 Franco-Prussian War, for example, France set up a series of programs intended to care for pregnant women, promote breastfeeding, and improve infant welfare. (David Barker has suggested, half seriously, that this early attention to maternal health lies behind the famous "French paradox": the low levels of cardiovascular disease found among the French, despite a diet rich in foie gras and triple crème cheese. It's not French people's nightly glass of red wine that protects them from heart attacks, Barker argues, but the programs and clinics provided to pregnant women more than a hundred years ago.)

In the United States, the First World War accelerated the development of the maternal health movement, for reasons that were laid out with brutal candor by Josephine Baker, chief of the Division of Child Hygiene of New York. "It may seem like a cold-blooded thing to say, but someone ought to point out that the World War was a back-handed break for children," Baker wrote in 1939. "As more and more thousands of men were slaughtered every day, the belligerent nations, on whatever side, began to see that new human lives, which could grow up to replace brutally extinguished adult lives, were extremely valuable national assets." When a nation is at war, she concluded, "it must look to its future supplies of cannon fodder." So consistently have military concerns motivated care for pregnant women and infants, writes historian Debórah Dwork, that one could say that "war is good for babies."

Current research on fetal origins suggests that the martial-minded leaders of past eras were correct: strong adults begin with healthy pregnant women. With the help of this research, could we now expand our aim beyond cultivating good soldiers to cultivating good students, workers, family members, and citizens, through programs that improve prenatal conditions? Interventions during pregnancy have several reasons to recommend them. For one, they can be precisely targeted to women during the nine months of their pregnancies—a far more manageable task than convincing the entire population to eat well or exercise regularly or stop smoking (though we should keep trying to do that, too). Indeed, some fetal origins researchers have explicitly compared the impact of fetal experience to that of adult health behaviors and found them similar in magnitude. Cardiologist John Deanfield and his colleagues, writing in the journal *Circulation* in 2001, reported that low birth weight has an effect on the functioning of the blood vessels in later life that is "as great as the effects of smoking." Each one-kilogram decrease in birth weight, Deanfield estimates, leads to a reduction in the capacity of blood vessels equivalent to smoking twenty cigarettes a day for four and a half years. Imagine if some fraction of the funds spent on antismoking billboards and package labels and public service announcements (all worthy and necessary efforts in their own right) were devoted to promoting prenatal nutrition.

A second reason to focus on pregnancy: early interventions are almost always more cost effective than later ones, as Douglas Almond notes. Because of the virtuous cycle that can be initiated by an early, successful intervention—health building on health, strength building on strength—"investments targeting fetal health may have higher rates of return than more traditional investments, such as schooling," he ventures. Money that improves schooling is always well spent, of course. But perhaps it would be even better spent long before an individual reaches school age—before birth, when his body and brain are still forming. Thomas Miller, an economist and resident fellow at the National Enterprise Institute, a Washington think tank, says that the choices we face remind him of a 1970s advertising slogan for Fram oil filters: "You can pay me now, or you can pay me later." "Multiplier effects mean that early interventions can be much less expensive and much more effective," he explains. "So, we can pay to help pregnant women now, or pay more to help their offspring later."

And third, though we still have much to learn about how to ensure optimal intrauterine conditions during pregnancy, these will likely prove more malleable than our genetic material, given the disappointing results yielded so far by technologies like gene therapy. "In contrast to the challenge of changing the determinism of our genes, the intrauterine environment is much more potentially modifiable by a woman to promote the good future health of her child," Duane Alexander, the director of the National Institute of Child Health and Human Development, has noted. At the same time, epigenetic research is revealing that intervening to make pregnancies healthier—or failing to do so—can have lasting effects that once were associated only with changes in genes. In a startling experiment published in

2005, scientist Michael Skinner exposed a pregnant rat to two commonly used industrial chemicals: vinclozolin, a fungicide, and methoxychlor, an insecticide. The rat's offspring had high rates of infertility, and were also prone to developing cancer, prostate disease, kidney disease, and immune cell defects.

The surprising, and alarming, thing was what happened next: When the offspring reproduced, *their* pups were also more likely to suffer from these conditions, as were the following two generations of rats—even though only the great-great-grandmother rat had had any direct encounter with the toxins. The effects of her exposure were being passed down to her descendants, not through changes to DNA but through changes to the epigenome. Indeed, a subsequent analysis by Skinner, director of Washington State University's Center for Reproductive Biology, identified changes in the expression of two genes associated with the health problems experienced by the rats. "These studies establish a novel mechanism of action not previously appreciated on how environmental toxins may act on a gestating mother to influence her grandchildren and subsequent generations," wrote Skinner and his coauthor, Matthew Anway. Skinner thinks it's likely that similar mechanisms are at work in human pregnancies and human diseases as well—meaning that the long-term impact of improving conditions for pregnant women and fetuses may be far greater than anyone has imagined.

What would such investments in enhancing the prenatal environment look like? The answers fairly spill out of the reporter's notebooks I've kept over the past eight months. We would improve access to prenatal care, and its quality. We would ensure that healthy, wholesome food was available to all pregnant women. We would make plans for protecting and provisioning pregnant women in emergency situations, such as natural disasters or terrorist attacks. We would offer pregnant women stress-reduction and support-building programs. We would regulate or ban chemicals that endanger fetuses, and would make more intensive efforts to understand how over-the-counter and prescription medications affect the fetus. We would provide drug treatment and smoking-cessation programs to pregnant women who are addicted. We would screen pregnant women for depression, anxiety, and other psychological conditions, and when problems were detected, we would offer therapy or counseling on the use of psychotropic medication during pregnancy.

Of course, many of these efforts are already being undertaken—but in a fragmented, incomplete fashion. If we've learned anything from fetal origins research, it's that all the factors impinging on a pregnancy work not in isolation but in interaction with each other, so that all must be addressed in order to promote the best possible outcome for the offspring. In the most optimistic scenario, such investments in the well-being of pregnant women and fetuses could result in a great upward leap in the population's health, akin to jumps seen in earlier eras following widespread improvements in nutrition or sanitation.

That is the sunny version of events, but there is a potential dark side, too. If history holds some hopeful signs that improving prenatal conditions can produce long-lasting benefits, it is also full of warnings about the dangers that lurk in such efforts. Consider, for example, a 1949 paper by Leo Kanner, an eminent physician known as the "father of child psychiatry" who was head of child psychiatry at Johns Hopkins Hospital. Kanner studied the roots of autism, and in an article in the *American Journal of Orthopsychiatry* he explained its cause: "Most of the patients were exposed from the beginning to parental coldness, obsessiveness, and a mechanical type of attention to material needs only. They were the objects of observation and experiment conducted with an eye on fractional performance rather than genuine warmth and enjoyment," Kanner wrote—and then came his most devastating line: "They were kept neatly in refrigerators which did not defrost." The "refrigerator mother" theory of autism, later promoted with great enthusiasm by the well-known psychologist Bruno Bettelheim, was just one instance in a long and virulent tradition of blaming mothers for anything that goes wrong with their children, from physical illness to juvenile delinquency, from excessive dependency to emotional unavailability. It's all too easy to imagine how fetal origins research could become the basis for a whole new species of mother-blame, finding fault with mothers even before their children are born.

By extension, fetal origins research could become the occasion to coerce or control the behavior of pregnant women. To see how readily this impulse may be roused we need look back only as far as the "crack panic" of the late 1980s, when reports of the use of crack cocaine by pregnant women in inner cities raised fears of a "bio-underclass, a generation of physically damaged cocaine babies whose biological inferiority is stamped at birth," in the words of the conservative columnist Charles Krauthammer, writing in the *Washington Post* in 1989. The nation's course of action was clear, Krauthammer continued: "We can either do nothing, or we can pass laws saying that any pregnant woman who takes cocaine during pregnancy will be sent until delivery to some not uncomfortable, secure location (boot camp, county jail, house arrest—the details are a purely technical matter) where she will be allowed everything except the liberty to leave or to take drugs." Needless to say, there is nothing in the discoveries of fetal origins research that changes pregnant women's status as citizens with rights, or as human beings with agency and free will.

Our growing awareness of the importance of maternal well-being to the fetus should lead us to offer help, not to force compliance or mete out punishment.

There's a final complication, perhaps most fraught of all: fetal origins research could become tangled in the ever-sticky web of abortion politics. When I ask fetal origins researchers about the relevance of their work to the debate over the morality and legality of abortion, they grow uneasy, shifting in their chairs and hesitating over their words. Fetal origins is concerned with the relationship between prenatal experience and postnatal life, they say at last, leaving me to complete the thought. For the aborted fetus, there is no postnatal life, so the matter of fetal origins is moot. That's true as far as it goes, but the contentious issue of abortion will not be so neatly contained. The findings of fetal origins research don't clearly favor the pro-life or the pro-choice faction; rather, they have the capacity to discomfit both sides, disturbing settled assumptions and undermining carefully negotiated distinctions.

Advocates of the right to abortion, for example, have at times portrayed the fetus as no more than an inert blob of tissue. Fetal origins research, with its emerging picture of a learning, adapting, responding fetus, makes that evasion less tenable. For their part, opponents of abortion have sometimes preferred to consider the fetus in splendid isolation, relegating the pregnant woman to the role of human incubator. Fetal origins research, with its elaboration of the intimate interplay between pregnant woman and fetus, does away with that imaginary separation. This point has been brought into sharp relief by recent research on the "wantedness" of a pregnancy and its effect on the offspring. Elizabeth Oltmans Ananat, an assistant professor of public policy and economics at Duke University, and Joanna Lahey, an assistant professor of public policy at Texas A&M University, used variation in nineteenth-century state laws against abortion as a natural experiment to test the long-term effects of being the product of an undesired pregnancy. They found that the "unwanted" child who was born due to abortion restrictions was as much as 50 percent less likely to survive to old age than was the average child born in the same era, perhaps because of low maternal investment in the pregnancy.

Contemplating all this, I find that I'm shifting uneasily in my chair, too; abortion is never a comfortable subject, least of all for a woman eight months pregnant with a fiercely wanted child. I find my mind straying to a memory almost twenty years old: I was in college, and my roommates and I had piled into a chartered bus that took us to Washington, D.C., for an abortion-rights rally. Marching arm in arm down a wide avenue cleared of traffic, I was stopped in my tracks by a picket sign held by a pro-life protester. It showed an enormous photograph of a fetus (probably one of Lennart Nilsson's images), next to the words, "You've Forgotten Someone." Too often, abortion-rights supporters, a group in which I continue to include myself, have indeed forgotten the fetus. But the pro-life forces have also "forgotten someone": the pregnant woman. Perhaps fetal origins research can help restore to the debate both parts of the whole.

If the influence of fetal origins research on the abortion debate is still murky, the effect in the other direction is clear, and troubling: the radioactive politics of abortion has led many scientists to avoid doing work, as one of them told me, "on anything involving the word 'fetus.'" Feelings about abortion run especially high in the United States, and they seem to have impeded the expansion of research into fetal origins here—a state of affairs I lament each time I dial, yet again, a long string of numbers on my phone, calling scientists in Britain, Canada, Australia, New Zealand, the Netherlands. To allow America's abortion politics to sideline this promising avenue of scientific inquiry would be an egregious mistake—a loss most of all for pregnant women and their fetuses, who stand to benefit the most from the field's discoveries.

Douglas Almond, for one, is undeterred. "Even though my work on fetal origins puts me on the margins of traditional economics, I can't get my mind off it," he confessed to me during my visit to his office. "Something just keeps pulling me back." An awareness of the importance of early-life experience has even led him to change his behavior in his personal life. When he traveled to China on a Fulbright scholarship in 2005, he was taken aback by the country's poor air quality. "The pollution in Beijing was so awful that you could see the soot in the air," he said. "I went running once and had a cough for days afterward." Almond's wife, Lena, was pregnant at the time. Concerned about how the pollution might affect their unborn child, they decided that she would visit him for no longer than a week. Now, Almond is examining the effects of coal burning on air quality in China, taking advantage of a convenient experiment of nature: the Chinese government's policy of providing free coal to homes and offices, but only in regions north of the Huai River. This policy has produced heavy air pollution in northern China—pollution that contributes to one of the highest rates of birth defects in the world.

Such natural experiments have, in large part, made the study of fetal origins possible. It's important to recognize, however, that many of these well-studied circumstances—from the Dutch Hunger Winter, to China's Great Leap Forward, to the terrorist attacks of 9/11—were anything but natural. They were the product of people's cruelty or grandiosity or fanaticism. As one fetal origins researcher has noted, ruefully, studies like his are "extracting otherwise inaccessible scientific knowledge from the harsh soil of human catastrophe." From this harsh soil we ought also to extract the knowledge that will allow us to prevent catastrophes from occurring in the future—whether that means providing adequate nutrition to disadvantaged pregnant women, or regulating environmental toxins that are harmful to fetuses, or protecting pregnant women in the wake of disasters. Above all, we must recognize that the health and well-being of fetuses is a matter of concern for our entire society, not simply of women who happen to be expecting a child. Our intense, and sometimes prurient or exploitative, interest in pregnancy needs to be matched by an actual investment in pregnancy, by a broad-based commitment to improving the conditions under which pregnancies happen.

This is a view shared by one of the world's most distinguished economists, the Harvard professor Amartya Sen. Sen won a Nobel Prize in 1998 for his work on the causes of global poverty, including studies on the economic impact of gender inequality. When women are deprived of education and opportunities, Sen has found, the whole society suffers. "The limited role of women's active agency seriously afflicts the lives of all people—men as well as women, children as well as adults," he wrote in his 1999 book, *Development as Freedom*. When women are empowered, his evidence shows, children are better educated, families are healthier, household incomes are higher, and civil society is stronger.

In recent years, Sen has turned his attention to another of what he calls the "hidden penalties of gender inequality": the effects of pregnant women's disadvantage on their fetuses. "The time has come to pay more attention to those consequences of gender inequality that have not yet received much attention, and indeed, have not yet been at all adequately investigated," Sen declared in a 2003 article he coauthored with fellow economist Siddiq Osmani. "These include the interconnections between gender inequality and maternal deprivation, on the one hand, and the health of children (of both sexes) and of adults that they grow into, on the other. Our argument is that women's deprivation in terms of nutrition and healthcare rebounds on the society in the form of ill-health of their offspring—males and females alike."

Sen, born in Santiniketan, India, offers an example from his native region: the exceptionally high levels of cardiovascular disease that have accompanied the recent rise in living standards in South Asia, he writes, can be explained by the fetal origins hypothesis. "Many of those who have managed to escape poverty within a single generation were born as low birth weight babies, because their mothers belonged to a generation that was steeped in poverty and mired in a culture of gender inequality," he notes. "They are the ones who now find themselves highly susceptible to the diseases of the new regime," such as heart disease and diabetes.

Although the neglect of women's needs "rebounds" on both sexes, Sen sees a special irony in its effect on men. "If I were religious, I would call it divine retribution," he told an interviewer from the *Times of India* in 2003. "We in India treat our women so badly that most are undernourished and give birth to underweight babies. It is known that cardiovascular diseases hit men much more than women. So, when you mistreat your women, the men eventually suffer."

I'm still mulling over Sen's words a few days later as I gingerly descend the stairs of the subway station in Union Square, loaded down with bags from the farmers market. As a train pulls up in a rush of wind and noise, I think of a scene described by the novelist and former *New York Times* columnist Anna Quindlen. When she was in *her* eighth month of pregnancy, she found herself pressed on all sides by a surging crowd on a subway platform. "As I looked around I saw that I was being surrounded by four women, some armed with shoulder bags," Quindlen writes. "'You need protection,' one said, and being New Yorkers, they ignored the fact that they did not know one another and joined forces to form a kind of phalanx around me, not unlike those that offensive linemen build around a quarterback. When the train arrived and the doors opened, they moved forward, with purpose, and I was swept inside, not the least bit bruised."

None of that is happening for me today. A long winter coat hides my belly; no one notices I'm pregnant. It's rush hour, and riders are pushing and shoving just to get inside the subway car. I manage to squeeze on with my bags, but I haven't come close to getting a seat when the doors close and the train starts to move. Suddenly the train lurches forcefully to a halt, and the car goes dark. When light and movement return a few moments later, I find myself sitting, stupidly, on the floor of the subway car, with apples rolling like marbles down the aisle and under the seats. I look around helplessly, feeling like a child lost in a thicket of pant legs and skirt hems.

Then, one set of hands after another drops an apple into my bag. From the far end of the car, a piece of my fruit is handed from one rider to another until it reaches me. Another set of hands pulls me up and gently guides me to a seat. In that moment I don't feel gawked or gaped at, embarrassed or self-conscious: just cared for, and grateful.

NINE MONTHS

It's sunrise on a morning in early March, and I am staring, dumb-founded, out the window of my apartment. The ground is blan-keted with snow, eight inches of it, and there's more coming down. In a few hours, if all goes as planned, my fetus will emerge into the world as a baby. But already all has not gone as planned. I hadn't anticipated this late-season blizzard, and now I'm wonder-ing if we'll be able to traverse the ninety blocks to the hospital in time for my scheduled caesarean section.

Turning away from the window, my eye is caught, as it often is, by a photograph on the wall above my kitchen table. It's a pic-ture of my first son taken a few minutes after he was born. His knit cap is skewed at a rakish angle, and he's wrapped in one of those standard-issue hospital blankets that make all newborns look like cousins. His eyes are squeezed tightly shut, as if he isn't yet ready to look at this new world he's landed in. In his tiny fea-tures I can see the trace of his three-year-old self, still tucked up in bed.

I never tire of this picture; I often find myself studying it as

I sip my tea in the morning, starting the day by contemplating another kind of beginning. Why do we find these first pictures so fascinating, why do we clamor for a photograph of mother and baby following a birth? To welcome a new life, yes—but also to witness a break, to see the separation between two individuals who until now have never been separated. This morning I feel full of excitement, and also of trepidation, an apprehension that is only partly about the unplowed streets outside. Birth is an ending as well as a beginning; it ruptures one relationship, even as it commences another. In a few hours I will no longer be pregnant— if only we can make it to the hospital.

It was birth that ruptured the relationship between Otto Rank and Sigmund Freud, the father of psychoanalysis. Rank was Freud's favored protégé, almost an adopted son. His family sent him to trade school to train as a locksmith, but at age nineteen Rank apprenticed himself to Freud instead, learning how to pick the lock of the unconscious. Soon Rank was a practicing psychoanalyst, with patients and (the source of the trouble that followed) ideas of his own.

In 1923, Rank became convinced that he had stumbled upon a great insight, which he began feverishly dictating to a secretary. Our psychological problems as adults, he announced, stem from the separation from our mothers we experienced as fetuses. From our time in the womb we retain a "memory of paradise," a haven from which we were rudely expelled; neurosis is "the unconscious reproduction of the anxiety at birth." The analytic situation re-creates the "intrauterine state" (Rank likened the patient's posture on the couch to the fetal position), "allowing the patient to repeat with better success in the analysis the separation from the mother." Within months, Rank had completed a book on the importance of what he called birth trauma, and he proudly sent the work—dedicated to Freud—to his mentor. But Freud's reaction was not what Rank had hoped. At the urging of Rank's rivals within the viperish circle of European analysts, Rank was tossed out of Freud's good graces and out of the psychoanalytic establishment. By diverging from Freud's orthodoxy, Rank had ensured his own expulsion from paradise.

Despite his exile, Rank's ideas took hold among his followers; in 1928, psychiatrist Marion Kenworthy wrote that the baby delivered by caesarean section "is prone to be less sensitized—he cries less, is markedly less irritated by the contacts of handling, etc.—than the child delivered through the birth canal." Kenworthy urged obstetricians to put their pregnant patients on diets in order to produce smaller babies, who would presumably experience less psyche-twisting trauma on their way down the birth canal. Over time, however, the idea Rank called "a matter of fundamental import" faded into a historical artifact. Then something surprising happened. In recent years, scientists have begun investigating a possibility akin to Rank's inspiration: that the experience of childbirth can exert a lasting influence on the infant.

"There is increasing evidence that perinatal stress and pain can have long-term effects—and the most stressful thing that happens to any baby is being born," says Vivette Glover, professor of perinatal psychobiology at Imperial College London. In 2001, she and her colleagues conducted a study that examined the level of stress hormones found in babies' umbilical cords immediately after birth. All experiences of childbirth prompt a flow of stress hormones in the fetus, but, Glover found, the level of these hormones varied by the type of delivery: assisted deliveries (those involving forceps or suction) produced the highest levels, caesarean sections the lowest, with normal vaginal deliveries located in between.

A year earlier, Glover had published a related study in the *Lancet*. In that experiment, she and her coauthors measured the rise in stress hormones and the intensity of crying in two-month-old infants after they'd received a routine inoculation shot. They found that the babies' responses to the shot were related to their mode of delivery, with the most vehement reaction found in those born by assisted delivery and the mildest response in those born by caesarean section. "It seems that, at least for the first eight weeks, the method of delivery affects the behavior and stress response of the baby," Glover tells me. "It may well have a more long-term effect, so that's the next thing we need to study."

Glover's findings are supported by research on the effects of pain experienced by infants *after* birth. Anna Taddio, a pain specialist at the Hospital for Sick Children in Toronto, noticed more than a decade ago that the male infants she treated seemed more sensitive to pain than their female counterparts. This discrepancy, she reasoned, could be due to sex hormones, to anatomical differences—or to a painful event experienced in this part of the world by many boys and no girls: circumcision. In a study of eighty-seven baby boys, Taddio found that those who had been circumcised soon after birth reacted more strongly and cried for longer than uncircumcised boys when they received a vaccination shot four to six months later. Among the circumcised boys, those who had received an analgesic cream at the time of the surgery cried less while getting the immunization than those circumcised without pain relief.

Taddio concluded that a single painful event could produce effects lasting for months, and perhaps much longer. "When we do something to a baby that is not an expected part of its normal development, especially at a very early stage, we may actually change the way the nervous system is wired," she says. Early encounters with pain may alter the threshold at which pain is felt later on, making a child hypersensitive to pain, or, alternatively, dangerously indifferent to it. Lasting effects might also include emotional and behavioral problems like anxiety and depression, even learning disabilities, though these findings are far more tentative. Some doctors who treat premature babies have begun to incorporate such findings into their practice, reducing the number of heel pricks to which the infants are subject, for example, and making their environments quieter, dimmer, and more womblike.

Could such long-term effects of pain apply to fetuses as well? Some researchers say it's possible, noting that the stress of an assisted delivery may well fall outside the range of a fetus's expected experience. (Forceps place up to fifty pounds of pressure on a fetus's head, as compared to the nineteen to thirty-three pounds of pressure exerted by the mother's tissues during a normal vaginal birth.) A few have gone so far as to suggest administering pain relief to fetuses undergoing an assisted birth, perhaps in the form of an injection of local anesthetic into the fetus's scalp where it is grasped by the forceps or suction device.

Vivette Glover speculates that even a normal birth may be painful for the fetus. "Given the enormous pressures exerted on the fetus by the maternal tissues, and given the general agreement that the pain-sensing system is in place and functioning in the fetus at forty weeks' gestation, it's puzzling that there has been so little consideration of the pain a fetus may feel while being born," says Glover. "Some say it's unlikely that the baby feels pain because childbirth is a natural process—but the experience of childbirth is natural for women as well, and many of them seek pain relief." Indeed, in the United States upwards of 87 percent of pregnant women receive some kind of medication for pain during labor.

In any case, as women and their doctors increasingly seek to manage all aspects of pregnancy and childbirth, there's less and less about the process that is entirely natural. Take the timing of birth, for example. For centuries, midwives and traditional healers have attempted to hasten or hold off the birth of a child; today, with the array of surgical and pharmaceutical interventions available, it can seem as if the long-sought power to choose when a baby will be born is now in doctors' hands. In the aggregate, such interventions can have surprising effects. Fewer babies are now born on Saturday or Sunday, for example. A 2003 study of more than 1.6 million births in California, published in the *Journal of the American Medical Association*, found that there were 17.5 percent fewer births on the weekend than expected, at least in part because doctors are more likely to schedule caesarean sections on weekdays.

Even the length of pregnancy itself has changed, partly as a result of medical interventions. In the decade between 1992 and 2002, the average duration of a pregnancy in the United States dropped from forty weeks to thirty-nine weeks, according to an analysis conducted by the March of Dimes. More American women are giving birth with the assistance of a surgeon: in 2005, the latest year for which data was available, 30.2 percent of all live births in the United States were caesarean deliveries, the highest rate ever reported. Women undergoing a scheduled C-section, like me, may know their future child's birthday months in advance. When I learned the day of the operation back in January, I jotted it in my datebook as if it were a hairdresser appointment or a dental checkup: "Baby born today!"

But nature still holds the power to upset our plans. Pulling the seatbelt tight across my belly, I take a deep breath as John and I set off for the hospital, moving across 110th Street at the magisterial pace of a parade float. Around us the world is white, the traffic-light smears of red and yellow and green the only colors. Turning onto First Avenue, we have the whole wide boulevard nearly to ourselves. I've never seen the city so still. John and I are quiet too: thinking, looking. Down a side street, I see a car fishtail as it pulls from its parking space; its back end swings out in wide arcs, an almost balletic motion, until its wheels gain traction and it moves off down the street. With a sudden loud clatter a snowplow hurtles past us, the edge of its plow striking sparks on manhole covers. At last the long sprawl of the hospital complex comes into view, and we pull up, right on time.

Perhaps because it was decided by fate for so long, the timing of one's birth—the month, the day, even the minute—has been imbued with great meaning in a great many cultures. Astrology is a belief as widespread and entrenched as it is entirely debunked. (I know this, and yet I'll admit that when my obstetrician consulted her calendar and offered me a delivery date in early March, my first thought was: a Pisces! I liked the idea of my fetus as an astrological fish, swimming ably in his own amniotic sea.) But like Otto Rank's notion of birth trauma, the idea that the time we are born may tell something about us is enjoying a surprising renaissance. Research on the "season of birth effect" looks for relationships between the timing of birth and outcomes later in life, such as the development of disease. In recent years such effects have been investigated by many researchers examining many different conditions, yielding a mass of varying and sometimes conflicting findings. Still, scientists have repeatedly found that people's season of birth may hold clues to their health decades later.

Studies of Union Army records from the Civil War, for example, have documented that soldiers born in spring and summer generally died earlier than those born in fall or winter. MIT economist Dora Costa has found that recruits with spring or summer birthdays were more likely to develop heart disease, and were 70 percent more likely to die of stroke, than those born later in the year. Costa suggests that is because their mothers had less to eat during the winter, and were more likely to have respiratory ailments while pregnant. But the news is not all bad for summer babies: in findings published in the *Journal of Clinical Endocrinology and Metabolism*, researchers at Bristol University in Britain found that, on average, children born in late summer or early fall are a centimeter taller and have thicker bones by age ten than their peers born at other times of the year—perhaps because their mothers were exposed during their pregnancies to more sunshine, which promotes the body's manufacture of bone-building vitamin D.

Season of birth effects have been reported for multiple sclerosis, epilepsy, autism, substance abuse, eating disorders, depression, even the timing of menopause and the risk of suicide. In most of these studies, the impact of the season of birth is very small. One of the strongest effects is found in schizophrenia; dozens of studies have concluded that schizophrenics are about 10 percent more likely than the rest of the population to have been born in the late winter and early spring, perhaps because of prenatal exposure to the viral infections more common at that time of year.

Although these studies focus on the timing of birth and its relationship to later-life outcomes, the influences that produce such outcomes seem to be acting *before* birth. In many cases, the precise mechanism is not known, and in others, the suspected influence is no longer relevant. Fresh food, for example, is now available all year round, meaning that winter is no longer a season of deprivation; Dora Costa finds no association between spring and summer birthdays and higher mortality rates in the generations born after the Civil War. Still, this body of research points once again to the notion that humans can be shaped before birth by external forces—the hours of sunshine in a day, the abundance and variety of food on the plate, the number and virulence of viruses in the air—and that these forces may even follow a seasonal calendar. The world has its way with us long before we're born.

No one has argued this point longer or more persuasively than Janet DiPietro, a professor of developmental psychology at Johns Hopkins University who many in the field regard as the dean of fetal studies. A few weeks before my due date, against my obstetrician's advice, I take a train to Baltimore to meet with DiPietro at her office at the university's Bloomberg School of Public Health. DiPietro has tousled auburn hair, an infectious laugh, and the comfortable air of a mother of three teenage boys, which she is. She describes to me how she came to work in this field.

"I was interested in why individuals are different from one another, in how a person's nervous system shapes who they are. I started out studying children, but I got worried about all the environmental influences that could be contaminating the results," she says. "So, I moved back in time, to study infants—but well, okay, babies experience environmental influence, too. So I moved back in time again, to study fetuses. What I came to realize is that the fetus has the biggest environmental contaminant of all, which is maternal physiology. You are never more closely integrated with your environment than when you are a fetus. So really, the joke is on me." But DiPietro doesn't for a minute regret her focus on the intrauterine milieu: "If I had my druthers, I'd study this for the rest of my life. I think it's the place with the most potential for true discovery."

The study of fetuses has often followed the same trajectory as DiPietro's career: ideas and approaches from child and infant psychology migrate back into the womb. A recently designed system for assessing the development of the fetus, for example, is modeled on one used to evaluate the newborn; the newborn is observed directly, while the fetus's behavior is viewed via ultrasound, but both are appraised on many of the same dimensions. DiPietro's current work with fetuses could also be seen as an application of a concept from child psychology—one that shook the foundations of the field when it was introduced in 1968. That year, a University of Virginia professor named Richard Bell published a paper containing a very simple observation. For decades, child psychologists had devoted unstinting efforts to showing how the parent influences the child: a punitive father produces a rebellious son; a loving mother turns out a well-adjusted daughter. The direction of influence seemed so clear as to be self-evident. But, Bell pointed out, the arrow could just as well be pointed the other way: a difficult, defiant boy could bring out the parental disciplinarian, while a placid, easygoing girl could evoke a mother's gentle affection. Our one-sided focus on the environment the parent provides for the child is "illogical," Bell observed in a later article, "because it causes us to overlook the fact that the child is a potent part of the environment for the parent!"

As hard as it may be to imagine, this concept, too, is being applied to the prenatal period. It's not only that the pregnant woman affects the fetus, through the many avenues explored in this book; the fetus, in turn, exerts an influence on the woman. I'd caught allusions to this idea in my research—the suggestion by scientists at the Harvard School of Public Health, for example, that women carrying boys have bigger appetites because of testosterone secreted by their fetuses. In a conference I attended on cutting-edge fetal research, a scientist spoke excitedly about what she and her colleagues called "maternal programming": the possibility that the physiological changes of pregnancy prime a woman's brain for parenting, that pregnancy grows not just a baby but a mother. Now, DiPietro tells me about how she came to investigate the possibility of maternal-fetal "interactionism": by accident.

"I was working with a statistician to do a second-by-second analysis of data I'd collected on pregnant women and their fetuses, expecting to find that changes in the mother's nervous system affected her fetus," DiPietro explains. "The statistician and I were looking over his work together, and he said, 'Isn't it so interesting how the fetus affects the mother!' And I said, 'You did the analysis backward. What you're talking about isn't possible.' But he insisted, and he was right: there was a relationship there, and it flowed from fetus to mother. Every time the fetus moved, the mother was getting a jolt to her nervous system, all below her conscious awareness." That was in 2003; now, DiPietro tells me, "the most exciting and important new strand in fetal studies is the idea of bidirectionality, the notion that influences on the pregnant woman and fetus go both ways." She invites me to visit her fetal assessment laboratory in a building across the street, where a study of such bidirectional effects is underway.

The walls of DiPietro's lab are papered with hundreds of snapshots of babies, a gallery of wide eyes and gummy smiles contributed by the mothers of former subjects. Over the course of almost two decades, DiPietro has collected data on more than a thousand fetuses. Her partner in this work is Kathleen Costigan, an obstetric nurse who shares DiPietro's friendly warmth;

as Costigan readies her equipment for the experiment, the two women chat about a Broadway show they recently attended. Soon the afternoon's first subject arrives; she's an employee here at the hospital, thirty weeks pregnant with her second child. DiPietro takes a seat off to the side, and Costigan helps the woman up onto the gurney, where she will get a quick ultrasound to identify the fetus's position. We all turn our heads toward the monitor, studying the image of the woman's fetus.

"He's going to have a big chin, like his older brother," the study subject says.

"And big feet!" exclaims Costigan. "A future size thirteen!"

"I've been keenly aware of those feet," the woman replies ruefully, rubbing her belly.

Ultrasound finished, Costigan places a probe on the woman's chest, and two on her abdomen; these will register the heart rates of the woman and her fetus, and how much the fetus is moving. Costigan's subject has done this before, and she knows the drill, holding up two fingers for the device that measures skin conductance, an indicator of nervous system activation. Next to the gurney, a boxy monitor begins spewing out a continuous sheet of paper. Maternal and fetal heart rates, fetal movement, and maternal skin conductance are traced on the paper in jagged lines like a seismograph. Costigan fiddles with a knob, and the galloping sound of a fetal heartbeat fills the room. I notice DiPietro looking at me.

"Is your baby moving around?" she asks. I realize that I've put a hand to my belly.

"Yes, he is," I say, surprised.

"He's probably reacting to the sound of the heartbeat," she says. She chuckles: "It's fetus-to-fetus communication."

This is not the first time my roles as a science writer and a pregnant woman have converged. Over the past nine months, the two have taken turns at the helm of my quest to understand fetal origins; after a while it was impossible to keep them separate, as I let my belly lead the way into labs, or felt my fetus kick, as if for emphasis, during interviews, or balanced my laptop on my knees while I reached around my ever-expanding middle to type. My pregnancy changed the way scientists related to me, too. During interviews, they would gesture at my belly to make a point, or would use the word "you" instead of "the pregnant woman." Catching sight of me at professional meetings, they often looked momentarily stunned, as if a whale had shown up at a conference of marine biologists. At those same meetings, held in gray hotel conference rooms under fluorescent lights, my eyes were probably the only ones to prick with tears when ultrasound images of fetuses or recorded wails of newborns punctuated the presentation of research findings. What was professional became personal for me, as well as the other way around; the two are inseparable, just as the science of pregnancy is inseparable from its history and culture.

I pull my attention back to the fetal assessment lab, where Kathleen Costigan is placing a mask over her subject's eyes, and a pair of bulky headphones over her ears. The headphones will fill with the strains of a baroque concerto, blocking out any sounds from the experiment, and the mask will prevent her from observing Costigan's actions.

"We're trying to sneak up on the baby without alerting the mom," Costigan explains. After giving the woman and her fetus a few minutes to settle down, she positions the stimulus over the woman's abdomen: a metal cookie tin, filled with unpopped popcorn kernels. (The tins make such a perfect experimental tool that Costigan bought out a grocery store's entire supply, and distributed the cookies inside them to her pregnant subjects.)

Costigan gives the improvised rattle a vigorous shake directly over the woman's belly, producing a startlingly sharp sound. The racket makes no impression on the subject, whose features remain relaxed. "I think she's asleep," says DiPietro. A minute later, Costigan repeats the motion with an empty cookie tin, an action that functions as an experimental control. The woman dozes on.

The experiment over, Costigan pulls the paper from the monitor and shows it to me. At the moment of the loud noise, she points out, the fetus moved—and the woman's nervous system jumped, too. The two lines leap in unison, though the subject was blissfully unaware that her fetus was moving or that her body was responding. Costigan slips off the headphones and gently lifts the mask from the woman's eyes; she blinks in the light.

"What happened?" the subject asks.

In today's and in previous studies, DiPietro finds that this call and response happens over and over again during pregnancy: the fetus moves, and the pregnant woman's nervous system starts. During the second half of pregnancy, the fetus moves about once a minute. The pregnant woman may feel as few as 16 percent of these movements, yet her body is always reciprocating. DiPietro is now extending this research to find out whether maternal-fetal interactions predict maternal-baby interactions later on. If a woman's body is especially responsive to the movements of her thirty-week-old fetus, will she be especially responsive to the needs of her six-month-old child? DiPietro ushers the study subject out of the lab, and turns back to me.

"I have a theory about the function served by this little fetal-maternal dance," she says with an impish smile, gesturing to my belly. "I think the baby is training you to pay attention to him, to hear his cries, to get up in the middle of the night. He's saying, 'Get ready—here I come!'"

*　　*　　*

The operating room is cold, so frigid it feels like I should be able to see my breath on the air, like someone has left a window open to the March wind. I hunch on the edge of a steel table as an anesthesiologist slides a long needle into my spine. John, dressed in green scrubs, is pacing outside in the hallway. The delivery nurse, funny and frank, tells me that because of their tendency to faint, husbands are not allowed to see their wives receive an epidural. "That big needle comes out and they hit the decks," she says.

I lie down on the table, extending my arms out to the side, where the nurse straps them down—"the crucifix position," she jokes. She sees that I'm shivering, and slips a warming blanket over my chest and shoulders; it inflates with pumped-in air, like the Michelin Man. It's the oddest collection of sensations: I feel cold, and warm, and from my sternum down, nothing at all, just a heavy numbness. Much has been written about the joy and the agony of vaginal birth, much less about delivering by caesarean. As an overwhelmingly passive experience, it doesn't lend itself to triumphal storylines; as a coolly clinical one, it doesn't allow much room for the poetry of the senses. What it does do is make disconcertingly plain the thin line between life in there and life out here. In place of Otto Rank's deeply symbolic journey down the birth canal, there is just a slit, a few layers of skin and fat and muscle pulled aside like a curtain to reveal what was here with us all along.

John is allowed to enter now, and he takes a seat near my head at the top of the operating table. The nurse has placed a blue drape between us and the surgical team hovering around the mound of my belly, and it feels cozy up here, like when you pull the covers up over your head at a slumber party. From behind the drape, we hear the hospital staff's loose chatter turn clipped, hear the clink of instruments on the tray, smell the odor of cauterized flesh. I feel no pain, just a strange sensation of internal tugging and pulling, as if my body is a suitcase in which someone is rummaging for their shoes. Surprisingly soon, we hear the obstetrician say, "I'm going to take the baby out now." At her beckoning, John stands up to peer over the drape, looking like a spectator at a baseball game.

There's a moment of stillness. I realize that I'm holding my breath. Then—an eruption of noise and motion. In the din there's a sound I hear above all the others: a baby's cry, rising in indignation.

"He's beautiful!" says the nurse.

"He's huge!" says John. He is laughing, and crying, his eyes fixed on his new son, who is now being weighed ("Ten pounds!" John crows) on the scale across the room.

"I wish you could see this," he says, and says again.

But I did. Turning my head, I saw the reflection of the doctor in a window whitened by snow, saw her faint form reach deep into me and lift out my baby. His image was pale and ghostly, the child of my imagination these long nine months. Now, the nurse carries him around to my side of the drape, and I see him in the flesh. His skin is a vivid pink, his hair a dark slick, his eyes glistening with ointment. There's nothing more real than a baby.

When we hold our babies for the first time, we imagine them clean and new, unmarked by life, when in fact they have already been shaped by the world, and by us. It's a koan of parenthood, one worthy of long contemplation: We are meeting someone we know well for the very first time.

NOTES

ONE MONTH

5 *"the larval stage of human development"*: Peter Gluckman and Mark Hanson, *The Fetal Matrix: Evolution, Development and Disease* (Cambridge: Cambridge University Press, 2004), p. 213.

5 *"a staging period for well-being and disease in later life"*: Janet A. DiPietro, "Prenatal Development," *Encyclopedia of Infant and Early Childhood Development*, edited by Marshall M. Haith and Janette B. Benson (San Diego: Academic Press, 2008), p. 604.

6 *the ancient Greeks' belief that looking at statues and other works of art*: See, e.g., Jan Bondeson, *A Cabinet of Medical Curiosities* (Ithaca, NY: Cornell University Press, 1997), p. 145, and Wendy Doniger and Gregory Spinner, "Misconceptions: Female Imaginations and Male Fantasies in Parental Imprinting," *Daedalus* (1998), vol. 127, no. 1.

6 *the belief among eighteenth-century Britons that cravings experienced by a pregnant woman*: See, e.g., Julia Epstein, "The Pregnant Imagination, Fetal Rights, and Women's Bodies: A Historical Inquiry," *Yale Journal of Law and the Humanities* (1995), vol. 7, no. 1, and Philip K. Wilson, "Eighteenth-Century 'Monsters' and Nineteenth-Century 'Freaks': Reading the Maternally Marked Child," in *Literature and Medicine* (2002), vol. 21, no. 1.

10 *"Surely we are all out of the computation of our age"*: Sir Thomas Browne, *Religio Medici* (Boston: Ticknor and Fields, 1862), p. 77.

10 *"Yes! The history of a man for the nine months preceding his birth"*: Samuel Taylor Coleridge, *The Notebooks of Samuel Taylor Coleridge*, vol. 4, edited by Merton Christensen and Kathleen Coburn (London: Routledge, 1990), p. 423.

TWO MONTHS

14   *"Because people knew so little about the fetus"*: Author interview with Barbara Luke.

15   *"Pregnant women should above all else avoid repletion"*: Galen, quoted in Peter Garnsey, *Food and Society in Classical Antiquity* (Cambridge: Cambridge University Press, 1999), p. 102.

15   *generations of midwives and other traditional healers*: See, e.g., Paul Fieldhouse, *Food and Nutrition: Customs and Culture* (New York: Taylor and Francis, 1986), p. 52, and Pamela Goyan Kittler and Kathryn Sucher, *Food and Culture* (New York: Wadsworth, 2007), p. 382.

15   *Prochownick . . . was the first to publish a study*: See, e.g., Barbara Luke and Timothy R. B. Johnson, "Nutrition and Pregnancy: A Historical Perspective and Update," *Women's Health Issues* (1991), vol. 1, no. 4, and Barbara Luke and Louis G. Keith, *Principles and Practice of Maternal Nutrition* (Park Ridge, NJ: Parthenon, 1992), p. 3.

16   *"It was believed that the baby would be fine"*: Author interview with Barbara Abrams.

16   *"perfect parasite"*: See, e.g., Peter D. Gluckman and Mark A. Hanson, "Maternal Constraint of Fetal Growth and its Consequences," *Seminars in Fetal and Neonatal Medicine* (2004), vol. 9, no. 5, and Andrew M. Prentice and Gail R. Goldberg, "Energy Adaptations in Human Pregnancy: Limits and Long-Term Consequences," *American Journal of Clinical Nutrition* (2000), vol. 71, no. 5.

16   *"the coarser vegetables"*: G. H. Napheys, quoted in Peter W. Ward, *Birth Weight and Economic Growth: Women's Living Standards in the Industrializing West* (Chicago: University of Chicago Press, 1993), p. 27.

17   *nearly two-thirds of American women of childbearing age are overweight*: Kathleen M. Rasmussen, Patrick M. Catalano, and Ann L. Yaktine, "New Guidelines for Weight Gain During Pregnancy: What Obstetrician/ Gynecologists Should Know," *Current Opinion in Obstetrics and Gynecology* (2009), vol. 21, no. 6.

17   *one in five women who give birth in the United States is obese*: Shin Y. Kim and others, "Trends in Prepregnancy Obesity in Nine States, 1993–2003," *Obesity* (2007), vol. 15, no. 4.

17   *up to 73 percent of U.S. women fail to follow guidelines*: Rasmussen, Catalano, and Yaktine, "New Guidelines for Weight Gain During Pregnancy: What Obstetrician/Gynecologists Should Know."

17   *overweight pregnant women are more likely to experience birth complications*: See American College of Obstetricians and Gynecologists, "Obesity

in Pregnancy," *Obstetrics and Gynecology* (2005), vol. 106, no 3. See also Jared M. Baeten, Elizabeth A. Bukusi, and Mats Lambe, "Pregnancy Complications and Outcomes Among Overweight and Obese Nulliparous Women," *American Journal of Public Health* (2001), vol. 91, no. 3, and Marie I. Cedergren, "Maternal Morbid Obesity and the Risk of Adverse Pregnancy Outcome," *Obstetrics and Gynecology* (2004), vol. 103, no. 2.

17 *overweight pregnant women are more likely to require a C-section*: Joshua L. Weiss and others, "Obesity, Obstetric Complications and Cesarean Delivery Rate—A Population-Based Screening Study," *American Journal of Obstetrics and Gynecology* (2004), vol. 190, no. 4, and Thomas K. Young and Barbara Woodmansee, "Factors That Are Associated with Cesarean Delivery in a Large Private Practice: The Importance of Prepregnancy Body Mass Index and Weight Gain," *American Journal of Obstetrics and Gynecology* (2002), vol. 87, no. 2.

17 *the incidence of some defects was twice as high among the children of obese mothers*: D. Kim Waller et al., "Prepregnancy Obesity as a Risk Factor for Structural Birth Defects," *Archives of Pediatric and Adolescent Medicine* (2007), vol. 161, no. 8.

17 *women who were overweight but not obese had a 15 percent increased risk*: Suzanne M. Gilboa and others, "Association Between Prepregnancy Body Mass Index and Congenital Heart Defects," *American Journal of Obstetrics and Gynecology* (2010), vol. 202, no. 1.

18 *greater weight gain by a woman during pregnancy was associated with a heavier child*: Emily Oken and others, "Gestational Weight Gain and Child Adiposity at Age 3 Years," *American Journal of Obstetrics and Gynecology* (2007), vol. 196, no. 4.

18 *this relationship persists into the offspring's adolescence*: Emily Oken and others, "Maternal Gestational Weight Gain and Offspring Weight in Adolescence," *Obstetrics and Gynecology* (2008), vol. 112, no. 5.

18 *children gestated by women post-surgery were 52 percent less likely to be obese*: John G. Kral and others, "Large Maternal Weight Loss from Obesity Surgery Prevents Transmission of Obesity to Children Who Were Followed for 2 to 18 Years," *Pediatrics* (2006), vol. 118, no. 6.

18 *children born after their mothers lost weight had lower birth weights*: J. Smith and others, "Effects of Maternal Surgical Weight Loss in Mothers on Intergenerational Transmission of Obesity," *Journal of Clinical Endocrinology and Metabolism* (2009), vol. 94, no. 11.

19 *"the two groups of siblings are different, physiologically speaking"*: Author interview with John Kral.

20 *"fed groups of pregnant and lactating rats"*: Stéphanie A. Bayol, Samantha J.

Farrington, and Neil C. Stickland, "A Maternal 'Junk Food' Diet in Pregnancy and Lactation Promotes an Exacerbated Taste for 'Junk Food' and a Greater Propensity for Obesity in Rat Offspring," *British Journal of Nutrition* (2007), vol. 98, no. 4.

21 *"Don't Eat That, or Your Child May Grow Up Fat"*: Deborah Kotz, *U.S. News and World Report*, September 13, 2007.

21 *a group of pregnant women was asked to drink carrot juice*: Julie A. Mennella, Coren P. Jagnow, and Gary K. Beauchamp, "Prenatal and Postnatal Flavor Learning by Human Infants," *Pediatrics* (2001), vol. 107, no. 6.

21 *young animals prefer flavors they encountered during gestation and lactation*: For a review, see Julie A. Mennella, "Mother's Milk: A Medium for Early Flavor Experiences," *Journal of Human Lactation* (1995), vol. 1.

21 *young rabbits devour the aromatic juniper berries their mothers ate*: Agnes Bilkó, Vilmos Altbäcker, and Robyn Hudson, "Transmission of Food Preference in the Rabbit: The Means of Information Transfer," *Physiology and Behavior* (1994), vol. 56, no. 5.

21 *wild mice feast on fennel like that consumed by their dams*: Paola Valsecchi, Anna Moles, and Marisa Mainardi, "Does Mother's Diet Affect Food Selection of Weanling Wild Mice?" *Animal Behavior* (1993), vol. 46.

21 *baby lab rats love the chocolate, rum, and walnut flavors*: P. J. Capretta, J. T. Petersik, and D. J. Steward, "Acceptance of Novel Flavours Is Increased After Early Experience of Diverse Taste," *Nature* (1975), vol. 254.

21 *"Mothers are giving information to their offspring"*: Author interview with Julie Mennella.

22 *flavors as varied as garlic, curry, cumin, fenugreek, mint, and vanilla*: See, e.g., Catherine A. Forestell and Julie A. Mennella, "Food, Folklore, and Flavor Preference Development," in *Nutrition and Health: Handbook of Nutrition and Pregnancy*, edited by Carol J. Lammi-Keefe and others (Totowa, NJ: Humana Press, 2008); Julie A. Mennella and Gary K. Beauchamp, "The Human Infants' Response to Vanilla Flavors in Mother's Milk and Formula," *Infant Behavior and Development* (1996), vol. 19, no. 1; and Julie A. Mennella, Anthony Johnson, and Gary K. Beauchamp, "Garlic Ingestion by Pregnant Women Alters Odor of Amniotic Fluid," *Chemical Senses* (1994), vol. 20, no. 20.

22 *assessed the reactions of newborn babies to the smell of anise*: Benoist Schaal, Luc Marlier, and Robert Soussignan, "Human Foetuses Learn Odours from Their Pregnant Mother's Diet," *Chemical Senses* (2000), vol. 25, no. 6.

23 *German troops blockaded western Holland, turning away all shipments of food*: See, e.g., Henri A. Van Der Zee, *The Hunger Winter: Occupied Holland, 1944–1945* (Lincoln, NE: University of Nebraska Press, 1998); Sharman Apt Russell, *Hunger: An Unnatural History* (New York: Basic Books,

2005); and Walter Gratzer, *Terrors of the Table: The Curious History of Nutrition* (New York: Oxford University Press, 2006).

24 *"people whose mothers were pregnant during the siege have more obesity, more diabetes, and more heart disease"*: Author interview with Tessa Roseboom. See also Tessa J. Roseboom and others, "Effects of Prenatal Exposure to the Dutch Famine on Adult Disease in Later Life: An Overview," *Twin Research and Human Genetics* (2001), vol. 41, no. 5, and the website of the Dutch Famine Birth Cohort Study: www.dutchfamine.nl/index.htm.

24 *"a tragic experiment of opportunity"*: Zena Stein, *Famine and Human Development* (New York: Oxford University Press, 1975), p. 4.

25 *"We were amazed by these early results"*: David Barker, *The Best Start in Life* (London: Century, 2003), p. 31.

26 *similar results have been reported by dozens of other scientists, studying a variety of different populations*: See, e.g., Jonathan R. Seckl, "Glucocorticoid Programming of the Fetus: Adult Phenotypes and Molecular Mechanisms," *Molecular and Cellular Endocrinology* (2001), vol. 185, no. 1–2.

26 *"I was initially very skeptical of the idea of fetal origins"*: Author interview with Janet Rich-Edwards.

27 *"One explanation is that fetuses are making the best of a bad job"*: Author interview with David Barker.

28 *On a massive scale, that's what is happening now in developing countries like India*: See, e.g., Andrew M. Prentice, "The Emerging Epidemic of Obesity in Developing Countries," *International Journal of Epidemiology* (2006), vol. 35, no. 1, and Andrew M. Prentice and Sophie E. Moore, "Early Programming of Adult Diseases in Resource Poor Countries," *Archives of Disease in Childhood* (2005), vol. 90.

28 *a striking 2003 experiment*: Robert A. Waterland and Randy L. Jirtle, "Transposable Elements: Targets for Early Nutritional Effects on Epigenetic Gene Regulation," *Molecular and Cellular Biology* (2003), vol. 23, no. 15.

30 *"only a small proportion of women planning a pregnancy"*: Hazel M. Inskip and others, "Women's Compliance with Nutrition and Lifestyle Recommendations Before Pregnancy: General Population Cohort Study," *British Medical Journal* (2009), vol. 338, no. 481.

31 *"Minamata disease"*: See, e.g., Philippe Grandjean, "Late Insights into Early Origins of Disease," *Basic and Clinical Pharmacology and Toxicology* (2008), vol. 102, no. 2, and Sandra Steingraber, *Having Faith: An Ecologist's Journey to Motherhood* (Cambridge: Perseus, 2001), pp. 43–48.

32 *researchers in New Zealand reported*: Tord E. Kjellstrom et al., *Physical and Mental Development of Children with Prenatal Exposure to Mercury from Fish* (Solna, Sweden: National Swedish Environmental Protection Board, 1986).

32 *scientists working in the Faroe Islands off the coast of Denmark*: Philippe Grandjean and others, "Cognitive Deficit in 7-Year-Old Children with Prenatal Exposure to Methylmercury," *Neurotoxicology and Teratology* (1997), vol. 19, no. 6. See also Philippe Grandjean et al., "Cognitive Performance of Children Prenatally Exposed to 'Safe' Levels of Methylmercury," *Environmental Research* (1998), vol. 77, no. 2.

32 *a report, commissioned by Congress, assessing the dangers of mercury exposure*: Committee on the Toxicological Effects of Methylmercury, Board on Environmental Studies and Toxicology, Commission on Life Sciences, National Research Council, *Toxicological Effects of Methylmercury* (Washington, DC: National Academy Press, 2000).

32 *following the FDA advisory, pregnant women reduced their fish consumption*: Emily Oken and others, "Decline in Fish Consumption Among Pregnant Women After a National Mercury Advisory," *Obstetrics and Gynecology* (2003), vol. 102, no. 2.

33 *"Omega-3 fatty acids are crucial for the development of the central nervous system"*: Author interview with Joseph Hibbeln.

33 *low levels of seafood consumption during pregnancy was associated with increased risk*: Joseph R. Hibbeln et al., "Maternal Seafood Consumption in Pregnancy and Neurodevelopmental Outcomes in Childhood (ALSPAC Study): An Observational Cohort Study," *Lancet* (2007), vol. 369, no. 9561.

35 *a "national eating disorder"*: Michael Pollan, *The Omnivore's Dilemma: A Natural History of Four Meals* (New York: Penguin Press, 2006), p. 2.

37 *"Pregnancy is actually an ideal occasion for shaking up the way you think and feel about food"*: Author interview with Karin Michels.

37 *Michels feels so strongly about this that she wrote a book intended for pregnant women*: Karin B. Michels and Kristine Napier, *The Gift of Health: The Complete Pregnancy Diet for Your Baby's Wellness—From Birth Through Adulthood* (New York: Pocket, 2001).

38 *"a number of chemicals that are protective against cancer"*: Author interview with David Williams.

38 *the offspring of mice who ingest a phytochemical derived from cruciferous vegetables*: Zhen Yu and others, "Indole-3-Carbinol in the Maternal Diet Provides Chemoprotection for the Fetus Against Transplacental Carcinogenesis by the Polycyclic Aromatic Hydrocarbon Dibenzo[a,l]pyrene," *Carcinogenesis* (2006), vol. 27, no. 10.

38 *the offspring of mice who were given green tea during pregnancy*: David J. Castro and others, "Chemoprevention of Dibenzo[a,l]pyrene Transplacental Carcinogenesis in Mice Born to Mothers Administered Green Tea: Primary Role of Caffeine," *Carcinogenesis* (2008), vol. 29, no. 8.

40 *compared to pregnant women who ate less than one serving of chocolate a week*:

Elizabeth W. Triche and others, "Chocolate Consumption in Pregnancy and Reduced Likelihood of Preeclampsia," *Epidemiology* (2008), vol. 19, no. 3.

40 *mothers who ate chocolate every day during pregnancy*: Katri Räikkönen and others, "Sweet Babies: Chocolate Consumption During Pregnancy and Infant Temperament at Six Months," *Early Human Development* (2004), vol. 76, no. 2.

THREE MONTHS

41 *"It's not your imagination"*: Author interview with David Fessler.

42 *pregnant women are more likely than non-pregnant women to react with disgust*: Daniel M. T. Fessler, Serena J. Eng, and Carlos D. Navarrete, "Elevated Disgust Sensitivity in the First Trimester of Pregnancy: Evidence Supporting the Compensatory Prophylaxis Hypothesis," *Evolution and Human Behavior* (2005), vol. 26, no. 4.

42 *while pregnant, women become more xenophobic*: Carlos D. Navarrete, Daniel M. T. Fessler, and Serena J. Eng, "Elevated Ethnocentrism in the First Trimester of Pregnancy," *Evolution and Human Behavior* (2007), vol. 28, no. 1.

42 *pregnant women have a stronger preference than do nonpregnant women for vigorous, healthy-looking faces*: Ben C. Jones et al., "Menstrual Cycle, Pregnancy and Oral Contraceptive Use Alter Attraction to Apparent Health in Faces," *Proceedings of the Royal Society B: Biological Sciences* (2005), vol. 272, no. 1561.

44 *About 1,700 of these people were pregnant women*: Philip J. Landrigan, "Impacts on the Health of Children of the September 11 Attacks on the World Trade Center," testimony given to the Subcommittee on Clear Air, Wetlands, and Climate Change Committee on Environment and Public Works, United States Senate, February 11, 2002, available at http://epw.senate.gov/107th/Landrigan_021102.htm.

44 *"I was leading a meeting at the center when I got a call"*: Author interview with Rachel Yehuda.

45 *Yehuda has coauthored more than a dozen articles about its effects on survivors*: See, e.g., Sarah R. Brand and others, "The Effect of Maternal PTSD Following in Utero Trauma Exposure on Behavior and Temperament in the 9-Month-Old Infant," *Annals of the New York Academy of Science* (2006), vol. 1071, and Stephanie Engel and others, "Psychological Trauma Associated with the World Trade Center Attacks and Its Effect on Pregnancy Outcome," *Paediatric and Perinatal Epidemiology* (2005), vol. 19, no. 5.

46 *"the scar without the wound"*: Arthur A. Cohen, *The Tremendum: A Theological Interpretation of the Holocaust* (New York: Crossroads, 1981), p. 2.

46 *low baseline or basal levels of cortisol were a marker of vulnerability to PTSD*: See, e.g., Rachel Yehuda, "Biological Factors Associated with Susceptibility to PTSD," *Canadian Journal of Psychiatry* (1999), vol. 44, no. 1.

46 *the offspring of people with PTSD have low basal cortisol as well*: See, e.g., Rachel Yehuda et al., "Parental Posttraumatic Stress Disorder as a Vulnerability Factor for Low Cortisol Trait in Offspring of Holocaust Survivors," *Archives of General Psychiatry* (2007), vol. 64, no. 9.

46 *thirty-eight women who were pregnant when they were exposed to the World Trade Center attack*: Rachel Yehuda et al., "Transgenerational Effects of Posttraumatic Stress Disorder in Babies of Mothers Exposed to the World Trade Center Attacks During Pregnancy," *Journal of Clinical Endocrinology and Metabolism* (2005), vol. 90, no. 7.

47 *offspring were more likely to develop PTSD if their mothers, but not their fathers, had PTSD*: Rachel Yehuda and others, "Maternal, Not Paternal, PTSD Is Related to Increased Risk for PTSD in Offspring of Holocaust Survivors," *Journal of Psychiatric Research* (2008), vol. 42, no. 13.

47 *"What looks like pathology to us might actually be a useful adaptation"*: Author interview with Jonathan Seckl. See also Jonathan R. Seckl and Michael J. Meaney, "Glucocorticoid 'Programming' and PTSD Risk," *Annals of the New York Academy of Science* (2006), vol. 1071, and Andrei Novac and S. Huber-Schneider, "Acquired Vulnerability: Comorbidity in a Patient Population of Adult Offspring of Holocaust Survivors," *American Journal of Forensic Psychiatry* (1998), vol. 19.

48 *"We can speculate that extra vigilance and rapid shifts in attention could be adaptive"*: Nicole M. Talge, Charles Neal, and Vivette Glover, "Antenatal Maternal Stress and Long-Term Effects on Child Neurodevelopment: How and Why?" *Journal of Child Psychology and Psychiatry* (2007), vol. 48, no. 3–4.

48 *"Our house is suspended over a hillside, and the whole structure shook"*: Author interview with Curt Sandman.

48 *he identified forty women who had experienced the earthquake*: Laura M. Glynn et al., "When Stress Happens Matters: Effects of Earthquake Timing on Stress Responsivity in Pregnancy," *American Journal of Obstetrics and Gynecology* (2001), vol. 184, no. 4.

49 *extremely stressful events can affect pregnancy outcomes*: See, e.g., Ali S. Khashan et al., "Reduced Infant Birthweight Consequent Upon Maternal Exposure to Severe Life Events," *Psychosomatic Medicine* (2008), vol. 70, and Ali S. Khashan et al., "Rates of Preterm Birth Following Antenatal

Maternal Exposure to Severe Life Events: A Population-Based Cohort Study," *Human Reproduction* (2009), vol. 24, no. 2.

49 *birth outcomes of 301 pregnant women living in New Orleans and Baton Rouge*: Xu Xiong et al., "Exposure to Hurricane Katrina, Post-Traumatic Stress Disorder, and Birth Outcomes," *The American Journal of the Medical Sciences* (2008), vol. 336, no. 2.

49 *women who had experienced the death or serious illness of a relative*: Ali S. Khashan et al., "Higher Risk of Offspring Schizophrenia Following Antenatal Maternal Exposure to Severe Adverse Life Events," *Archives of General Psychiatry* (2008), vol. 65, no. 2.

50 *pregnant women with high levels of cortisol*: Miguel A. Diego et al., "Maternal Psychological Distress, Prenatal Cortisol, and Fetal Weight," *Psychosomatic Medicine* (2006), vol. 68, no. 5.

51 *167 children whose fathers died while they were still in the womb*: Matti O. Huttunen and Pekka Niskanen, "Prenatal Loss of Father and Psychiatric Disorders," *Archives of General Psychiatry* (1978), vol. 35, no. 4.

51 *the offspring of women who were in their second month of pregnancy during June of 1967*: Dolores Malaspina and others, "Acute Maternal Stress in Pregnancy and Schizophrenia in Offspring: A Cohort Prospective Study," *BMC Psychiatry* (2008), vol. 8, issue 71.

53 *"The ice froze on everything"*: Author interview with Suzanne King.

54 *the more stressful events pregnant women encountered during the disaster, the lower were their babies' birth weights*: Suzanne King and David P. Laplante, "The Effects of Prenatal Maternal Stress on Children's Cognitive Development: Project Ice Storm," *Stress* (2005), vol. 8, no. 1.

54 *an association between prenatal stress and cognitive and language skills*: David P. Laplante et al., "Stress During Pregnancy Affects General Intellectual and Language Functioning in Human Toddlers," *Pediatric Research* (2004), vol. 56, no. 3.

54 *continued cognitive and language delays*: David P. Laplante et al., "Project Ice Storm: Prenatal Maternal Stress Affects Cognitive and Linguistic Functioning in 5 ½-Year-Old Children," *Journal of the American Academy of Child and Adolescent Psychiatry* (2008), vol. 47, no 9.

55 *focused emergency planners on the need to care for victims who are pregnant*: See, e.g., William M. Callaghan et al., "Health Concerns of Women and Infants in Times of Natural Disasters: Lessons Learned from Hurricane Katrina," *Maternal and Child Health Journal* (2007), vol. 11, no. 4, and Jeanne Pfeiffer and others, "Maternal and Newborn Care During Disasters: Thinking Outside the Hospital Paradigm," *Nursing Clinics of North America* (2008), vol. 43, no. 3.

55 *Katrina alone displaced at least ten thousand pregnant women*: Cited in Pierre Buekens et al., "Guest Editorial: Hurricanes and Pregnancy," *Birth* (2006), vol. 33, no. 2.

56 *some researchers have advocated incorporating disaster planning into childbirth education classes*: See, e.g., Lauren Dewald and Lily Fountain, "Introducing Emergency Preparedness in Childbirth Education Classes," *Journal of Perinatal Education* (2006), vol. 15, no. 1.

57 *recommendations we find on the website of the American Red Cross*: "Three Steps to Preparedness," website of the American Red Cross, www.redcross .org.

58 *"when the mistress has perforce to contemplate an enforced retirement from public life"*: Jane Ellen Panton, *From Kitchen to Garret: Hints for Young Householders* (London: Walk and Downey, 1888), p. 180. See also Judith Flanders, *Inside the Victorian Home: A Portrait of Domestic Life in Victorian England* (New York: W.W. Norton, 2004), p. 51.

58 *more than two-thirds of American pregnant women work*: "Maternity Leave and Employment Patterns: 1961–2003," U.S. Census Bureau, 2008.

59 *"I know that when I was pregnant, I liked being macho"*: Author interview with Suzanne King.

60 *"a lot of people thought it was just ridiculous"*: Author interview with Calvin Hobel.

60 *enrolling more than six hundred working women in Ireland*: Isabelle Niedhammer et al., "Occupational Predictors of Pregnancy Outcomes in Irish Working Women in the Lifeways Cohort," *British Journal of Obstetrics and Gynecology* (2009), vol. 116, no. 7.

61 *identify and reduce sources of stress before pregnancy*: For more, see Calvin J. Hobel, Amy Goldstein, and Emily S. Barrett, "Psychosocial Stress and Pregnancy Outcome," *Clinical Obstetrics and Gynecology* (2008), vol. 51, no. 2.

61 *screened once a trimester for "psychosocial risk factors"*: ACOG Committee on Health Care for Underserved Women, "Psychosocial Risk Factors: Perinatal Screening and Intervention," *Obstetrics and Gynecology* (2006), vol. 108, no. 2.

62 *"Most pregnant women can stop worrying about stress"*: Author interview with Janet DiPietro.

62 *"mild stress may be a necessary condition for optimal development"*: Janet A. DiPietro, "The Role of Prenatal Maternal Stress in Child Development," *Current Directions in Psychological Science* (2004), vol. 13, no. 2.

62 *women who reported modest anxiety and daily stress during pregnancy*: Janet A. DiPietro and others, "Maternal Psychological Distress During Pregnancy in Relation to Child Development at Age Two," *Child Development* (2006), vol. 77, no. 3.

62 *direct biological evidence of this relationship between prenatal stress and accelerated development*: Janet A. DiPietro et al., "Prenatal Antecedents of Newborn Neurological Maturation," *Child Development* (2010), vol. 81, no. 1.

64 *"Women's Work Stress Harms Unborn Babies"*: Claire Masters, *Daily Telegraph*, May 15, 2006.

64 *"Pregnant Women in High Stress Jobs Should Cut Working Hours"*: Agence France Press, April 21, 2006.

64 *"Working 32 Hours a Week or More in Pregnancy 'Is As Risky as Smoking'"*: Nicola Thomas, *Sunday Telegraph*, May 28, 2006.

64 *women who experienced high "job strain" while pregnant*: Marcel F. van der Wal, Manon van Eijsden, and Gouke J. Bonsel, "Stress and Emotional Problems During Pregnancy and Excessive Infant Crying," *Journal of Developmental and Behavioral Pediatrics* (2007), vol. 28, no. 6.

64 *a high level of job strain, and a workweek of 32 hours or more*: Tanja G.M. Vrijkotte et al., "First-Trimester Working Conditions and Birthweight: A Prospective Cohort Study," *American Journal of Public Health* (2009), vol. 99, no. 8.

65 *the Brandeis Brief*: Louis D. Brandeis, brief for *Muller v. Oregon*, Supreme Court of the United States, 1907.

66 *"That woman's physical structure and the performance of maternal functions place her at a disadvantage"*: David Josiah Brewer, opinion in *Muller v. Oregon*, Supreme Court of the United States, 1908.

67 *"epoch in American civilization"*: Quoted in Robert D. Johnson, *The Radical Middle Class: Populist Democracy and the Question of Capitalism in Progressive Era Portland, Oregon* (Princeton, NJ: Princeton University Press, 2003), p. 21.

67 *"Does anybody doubt that the only way you can get work out of a horse"*: Louis D. Brandeis, brief for *Stettler v. O'Hara*, Supreme Court of the United States, 1916.

67 *"the cow-like tranquility of pregnant women"*: Vivienne Parry, "Women: The Sex with More Stress?" *The Times* of London, July 1, 2006.

68 *"fetal protection policies," which claimed concern for the fetus*: See, e.g., Mary E. Becker, "From *Muller v. Oregon* to Fetal Vulnerability Policies," *University of Chicago Law Review*, vol. 53, no. 4, and Juliana S. Gonen, "Women's Rights vs. 'Fetal Rights': Politics, Law and Reproductive Hazards in the Workplace," in *The Politics of Pregnancy: Policy Dilemmas in the Maternal-Fetal Relationship*, edited by Janna C. Merrick and Robert H. Blank (New York: Hayworth Press, 1993).

69 *"Concern for a woman's existing or potential offspring historically has been the excuse"*: Harry A. Blackmun, opinion in *Automobile Workers v. Johnson Controls*, Supreme Court of the United States, 1991.

71 *compared a group of sixty-two women who took antenatal leave to 385 women who worked up until their babies' births*: Sylvia Guendelman et al., "Maternity Leave in the Ninth Month of Pregnancy and Birth Outcomes Among Working Women," *Women's Health Issues* (2009), vol. 19, no. 1.

71 *"Women who take time off from work"*: Author interview with Sylvia Guendelman.

71 *relaxation exercises decreased participants' feelings of stress*: Jeronima Teixeira and others, "The Effects of Acute Relaxation on Indices of Anxiety During Pregnancy," *Journal of Psychosomatic Obstetrics and Gynecology* (2005), vol. 26, no. 4.

71 *several studies of prenatal yoga, all carried out in India*: See, e.g., Maharana Satyapriya et al., "Effect of Integrated Yoga on Stress and Heart Rate Variability in Pregnant Women," *International Journal of Gynaecology and Obstetrics* (2009), vol. 104, no. 3; Shamanthakamani Narendran et al., "Efficacy of Yoga on Pregnancy Outcome," *Journal of Alternative and Complementary Medicine* (2005), vol. 11, no. 2; and Shamanthakamani Narendran et al., "Efficacy of Yoga in Pregnant Women with Abnormal Doppler Study of Umbilical and Uterine Arteries," *Journal of the Indian Medical Association* (2005), vol. 103, no. 1.

FOUR MONTHS

77 *a stack of histories of pregnancy and alcohol*: See, e.g., Elizabeth Armstrong, *Conceiving Risk, Bearing Responsibility: Fetal Alcohol Syndrome and the Diagnosis of Moral Disorder* (Baltimore: Johns Hopkins University Press, 2003), and Janet Golden, *Message in a Bottle: The Making of Fetal Alcohol Syndrome* (Cambridge: Harvard University Press, 2005).

78 *maternity wards that smelled, recalled one doctor, "like a bar"*: Quoted in Armstrong, *Conceiving Risk, Bearing Responsibility*, p. 74.

78 *"At the better hospitals, women in labor might be handed a vodka and orange juice"*: Author interview with Janet Golden.

78 *"Alcohol, as such, is not injurious and need not be eliminated during pregnancy"*: Quoted in Armstrong, *Conceiving Risk, Bearing Responsibility*, p. 72.

78 *"It can be stated categorically that no matter how great the amounts of alcohol imbibed by the mother or the father"*: Ashley Montagu, *Life Before Birth* (New York: New American Library, 1964), pp. 100–101.

78 *"No, smoking and alcoholic drinks have no effect on an unborn baby"*: Quoted in Richard M. Dorson, "Folklore in the News: Pregnancy Superstitions," *Western Folklore* (1955), vol. 14, no. 1.

79 *"The Victorian tendency to put woman on a pedestal led to the idealization of the womb"*: Ann Dally, "Thalidomide: Was the Tragedy Preventable?" *Lancet* (1998), vol. 351, no. 9110.

79 *"the miracle drugs that have tumbled from the laboratories in such heartening profusion recently"*: Quoted in Robert Meyers, *D.E.S.: The Bitter Pill* (New York: Seaview/Putnam, 1983), p. 90.

79 *"were among the most medicated women in history"*: Meyers, *D.E.S.: The Bitter Pill*, p. 15.

80 *half of all new mothers took two to four pharmaceutical products while pregnant*: Cited in Meyers, *D.E.S.: The Bitter Pill*, p. 15.

80 *The first story begins in Germany on a summer day in 1961*: See, e.g., Geoff Adams-Spink, "Thalidomide: 40 Years On," BBC, June 7, 2002; F. Clarke Fraser, "Thalidomide Retrospective: What Did We Learn?" *Teratology*, vol. 38, no. 3; Phillip Knightley et al., *Suffer the Children: The Story of Thalidomide* (New York: Viking, 1979); Widukind Lenz, "A Short History of Thalidomide Embryopathy," *Teratology* (1988), vol. 38, no. 3; Widukind Lenz, "A Personal Perspective on the Thalidomide Tragedy," *Teratology* (1992), vol. 46, no. 5; William A. Silverman, "The Schizophrenic Career of a 'Monster Drug,'" *Pediatrics* (2002) vol. 110, no. 2; and Trent Stephens and Rock Brynner, *Dark Remedy: The Impact of Thalidomide and Its Revival as a Vital Medicine* (New York: Basic Books, 2001).

81 *"Every month's delay in clarification"*: Widukind Lenz, quoted in Stephens and Brynner, *Dark Remedy*, p. 31.

82 *The second story begins six years later, on Independence Day in Massachusetts*: See, e.g., Roberta J. Apfel and Susan M. Fisher, *To Do No Harm: DES and the Dilemmas of Modern Medicine* (New Haven: Yale University Press, 1986); Richard Gillam and Barton J. Bernstein, "Doing Harm: The DES Tragedy and Modern American Medicine," *The Public Historian* (1987), vol. 9, no. 1; Meyers, *D.E.S: The Bitter Pill*, 1983; Howard Ulfelder, "The Stilbestrol-Adenosis-Carcinoma Syndrome," *Cancer* (1976), vol. 38, no. 1; and Howard Ulfelder, "The Stilbestrol Disorders in Historical Perspective," *Cancer* (1980), vol. 45, no. 12.

82 *"There must be some explanation for this explosion"*: Howard Ulfelder, quoted in Meyers, *D.E.S.: The Bitter Pill*, p. 93.

82 *"I thought, 'My god . . . '"*: Howard Ulfelder, quoted in Meyers, *D.E.S.: The Bitter Pill*, p. 94.

82 *"You can stop your study," said a mother whose daughter had died of the cancer*: Quoted in Meyers, *D.E.S.: The Bitter Pill*, p. 94.

83 *the physicians had collected enough evidence to make an astonishing claim*: Arthur L Herbst, Howard Ulfelder, and David C. Poskanzer, "Adenocarcinoma of the Vagina: Association of Maternal Stilbestrol Therapy with Tumor

Appearance in Young Women," *New England Journal of Medicine* (1971), vol. 284, no. 15.

83   *DES had been prescribed to around four million American women*: Sarina Schrager and Beth E. Potter, "Diethylstilbestrol Exposure," *American Family Physician* (2004), vol. 69, no. 10.

83   *as many as eight million women worldwide*: Retha R. Newbold, "Prenatal Exposure to Diethylstilbestrol (DES)," *Fertility and Sterility* (2008), vol. 89, no. 2.

83   *Daughters exposed to DES in utero have forty times the risk*: Elizabeth E. Hatch and others, "Cancer Risk in Women Exposed to Diethylstilbestrol in Utero," *Journal of the American Medical Association* (1998), vol. 280, no. 7.

83   *the world's first transplacental carcinogen*: See, e.g., John A. McLachlan, "Prenatal Exposure to Diethylstilbestrol (DES): A Continuing Story," *International Journal of Epidemiology* (2006), vol. 35, no. 4, and Newbold, "Prenatal Exposure to Diethylstilbestrol (DES)."

83   *The third and final story began in early February, 1973*: See, e.g., Faye Calhoun and Kenneth Warren, "Fetal Alcohol Syndrome: Historical Perspectives," *Neuroscience and Biobehavioral Reviews* (2007), vol. 31, no. 2, and Kenneth L. Jones, David W. Smith, Christy N. Ulleland, and Ann Pytkowicz Streissguth, "Pattern of Malformation in Offspring of Chronic Alcoholic Mothers," *Lancet* (1973), vol. 1, no 7815.

84   *"At first we really doubted ourselves"*: Author interview with Kenneth Lyons Jones.

84   *such a relationship had been speculated about for centuries*: See, e.g., Ernest L. Abel, "Was the Fetal Alcohol Syndrome Recognized by the Greeks and Romans?" *Alcohol and Alcoholism* (1999), vol. 34, no. 6, and Lynn Martin, "Fetal Alcohol Syndrome in Europe, 1300–1700: A Review of Data on Alcohol Consumption and a Hypothesis," *Food and Foodways* (2003), vol. 11, no. 1.

84   *Jones and his colleagues in Seattle were the first to define a formal diagnosis*: Jones, Smith, Ulleland, and Streissguth, "Pattern of Malformation in Offspring of Chronic Alcoholic Mothers."

85   *"During most of pregnancy, the placenta separating mother and fetus is only one cell thick"*: Author interview with Gideon Koren.

86   *Thalidomide seems to wreak its devastation by inhibiting the growth of new blood vessels*: See, e.g., Trent D. Stephens, Carolyn J. W. Bunde, and Bradley J. Fillmore, "Mechanism of Action in Thalidomide Teratogenesis," *Biochemical Pharmacology* (2000), vol. 59, no. 12. For the most recent advance in the understanding of thalidomide's mechanism of action, see Takumi Ito and others, "Identification of a Primary Target of Thalidomide Teratogenicity," *Science* (2010), vol. 327, no. 5971.

86 *DES appears to impersonate one of the body's own hormonal messengers*: See, e.g., Hugh S. Taylor, "Endocrine Disruptors Affect Developmental Programming of HOX Gene Expression," *Fertility and Sterility* (2008), vol. 89, no. 2.

86 *alcohol may trigger the death of cells in the fetus's growing brain*: See, e.g., Chrysanthy Ikonomidou et al., "Ethanol-Induced Apoptotic Neurodegeneration and Fetal Alcohol Syndrome," *Science* (2000), vol. 287, no. 5455.

86 *many of the monsters of ancient Greek mythology . . . appear to be based on types of congenital malformations*: See, e.g., Gert-Horst Schumacher, "Teratology in Cultural Documents and Today," *Annals of Anatomy* (2004), vol. 186, no. 5–6.

87 *more than three-quarters took at least one medication during pregnancy*: Jerrie S. Refuerzo and others, "Use of Over-the-Counter Medications and Herbal Remedies in Pregnancy," *American Journal of Perinatology* (2005), vol. 22, no. 6.

87 *women take more over-the-counter drugs when they're pregnant*: Michele Y. Splinter and others, "Prenatal Use of Medications by Women Giving Birth at a University Hospital," *Southern Medical Journal* (1997), vol. 90, no. 5, and Martha M. Werler et al., "Use of Over-the-Counter Medications During Pregnancy," *American Journal of Obstetrics and Gynecology* (2005), vol. 193, no. 3.

87 *pregnant women are taking more such drugs now*: Werler, "Use of Over-the-Counter Medications During Pregnancy."

87 *pain relievers, cold remedies, and allergy medications*: Lisa McKenna and Meredith McIntyre, "What Over-the-Counter Preparations Are Pregnant Women Taking? A Literature Review," *Journal of Advanced Nursing* (2006), vol. 56, no. 6, and Jackie Tillett, Loryjean Medved Kostich, and Leona VandeVusse, "Use of Over-the-Counter Medications During Pregnancy," *Journal of Perinatal and Neonatal Nursing* (2003), vol. 17, no. 1.

87 *65 percent took acetaminophen . . . and 15 percent took pseudoephedrine*: Werler, "Use of Over-the-Counter Medications During Pregnancy."

87 *64 percent of women were prescribed at least one drug during pregnancy*: Susan E. Andrade and others, "Prescription Drug Use in Pregnancy," *American Journal of Obstetrics and Gynecology* (2004), vol. 191, no. 2.

88 *American women take an average of three to five prescription drugs*: The U.S. Food and Drug Administration, "FDA Proposes New Rule to Provide Updated Information on the Use of Prescription Drugs and Biological Products During Pregnancy and Breast-Feeding," news release, May 28, 2008.

88 *The FDA itself has acknowledged*: Food and Drug Administration, "FDA Proposes New Rule."

88 *79 percent of pregnant women take at least one drug for which safety data are unavailable*: Isabelle Lacroix et al., "Prescription of Drugs During Pregnancy in France," *Lancet* (2000), vol. 356, no. 9243.

88 *almost one-half of pregnant women were prescribed drugs labeled C, D, or X*: Andrade, "Prescription Drug Use in Pregnancy."

89 *16.4 percent of pregnant women smoke cigarettes*: Office of Applied Studies, "Results from the 2008 National Survey on Drug Use and Health: National Findings," Substance Abuse and Mental Health Services Administration, U.S. Department of Health and Human Services, available at http://www.oas.samhsa.gov/nsduh/2k8nsduh/2k8Results.cfm#2.6. See also Van T. Tong and others, "Trends in Smoking Before, During, and After Pregnancy: Pregnancy Risk Assessment Monitoring System (PRAMS), United States, 31 Sites, 2000–2005," *Morbidity and Mortality Weekly Report Surveillance Summary* (2009), vol. 58, no. 4.

89 *5.1 percent of pregnant women use an illicit drug*: Office of Applied Studies, "Results from the 2008 National Survey on Drug Use and Health: National Findings."

89 *some pregnant women drink alcohol, about one in five*: "Not Even One," HHS HealthBeat, U.S. Department of Health and Human Services, December 2, 2008, available at http://www.hhs.gov/news/healthbeat/2008/12/20081202a.html.

89 *one particularly sneaky study, conducted in Sweden*: Friedrich Martin Wurst and others, "Measurement of Direct Ethanol Metabolites Suggests Higher Rate of Alcohol Use Among Pregnant Women Than Found with the AUDIT—A Pilot Study in a Population-Based Sample of Swedish Women," *American Journal of Obstetrics and Gynecology* (2008), vol. 198, no. 4.

89 *problems with their offspring's brain development and behavior*: See, e.g., Monique Ernst, Eric T. Moolchan, and Miqun L. Robinson, "Behavioral and Neural Consequences of Prenatal Exposure to Nicotine," *Journal of the American Academy of Child and Adolescent Psychiatry* (2001), vol. 40, no. 6; Melissa Hermann, Katherine King, and Michael Weitzman, "Prenatal Tobacco Smoke and Postnatal Secondhand Smoke Exposure and Child Neurodevelopment," *Current Opinion in Pediatrics* (2008), vol. 20, no. 2; and Sabine J. Roza et al., "Maternal Smoking During Pregnancy and Child Behaviour Problems: The Generation R Study," *International Journal of Epidemiology*, vol. 38, no. 3.

89 *if they quit smoking before they reach week fifteen of their pregnancies*: Lesley M. McCowan et al., "Spontaneous Preterm Birth and Small for Gestational Age Infants in Women Who Stop Smoking Early in Pregnancy: Prospective Cohort Study," *British Medical Journal* (2009), vol. 338.

90 *The Maternal Lifestyle Study*: Henrietta S. Bada et al., "Impact of Prenatal Cocaine Exposure on Child Behavior Problems Through School Age," *Pedatrics* (2007), vol. 119, no. 2.

90 *a 2008 study of 12,500 three-year-olds*: Yvonne Kelly et al., "Light Drinking in Pregnancy, a Risk for Behavioural Problems and Cognitive Deficits at 3 Years of Age?" *International Journal of Epidemiology* (2009), vol. 38, no. 1.

91 *"it may also be that light-drinking mothers tend to be more relaxed"*: Yvonne Kelly, quoted in Catharine Paddock, "Children Not Harmed by Small Amount of Alcohol in Pregnancy," *Medical News Today* (2008).

92 *if they have either of two common genetic traits*: See, e.g., Xiaobin Wang et al., "Maternal Cigarette Smoking, Metabolic Gene Polymorphism, and Infant Birth Weight," *Journal of the American Medical Association* (2002), vol. 287, no. 2.

93 *71 percent of the babies of poor alcoholic women had fetal alcohol syndrome*: Nesrin Bingol et al., "The Influence of Socioeconomic Factors on the Occurrence of Fetal Alcohol Syndrome," *Advances in Alcohol and Substance Abuse* (1987), vol. 6, no. 4.

93 *"synthetic man"*: Quoted in Sarah A. Vogel, "From 'The Dose Makes the Poison' to 'The Timing Makes the Poison': Conceptualizing Risk in the Synthetic Age," *Environmental History* (2008), vol. 13, no. 4.

94 *the National Toxicology Program had released a preliminary report*: Center for the Evaluation of Risks to Human Reproduction, National Toxicology Program, U.S. Department of Health and Human Services, "Interim Draft: NTP-CERHR Report on the Reproductive and Developmental Toxicity of Bisphenol A," April 2007.

96 *first synthesized in 1938 by a British scientist, Charles Edward Dodds*: See, e.g., Apfel and Fisher, *To Do No Harm*, and Meyers, *D.E.S.: The Bitter Pill*.

97 *they wreak havoc by interfering with the body's natural hormonal signaling system*: See, e.g., Theo Colborn, Dianne Dumanoski, and John Peterson Myers, *Our Stolen Future: Are We Threatening Our Fertility, Intelligence, and Survival?: A Scientific Detective Story* (New York: Dutton, 1996), and Sheldon Krimsky, *Hormonal Chaos: The Scientific and Social Origins of the Environmental Endocrine Hypothesis* (Baltimore: Johns Hopkins University Press, 1999).

99 *hairdressers and beauticians frequently exposed during pregnancy to hairspray*: Gillian Ormond and others, "Endocrine Disruptors in the Workplace, Hair Spray, Folate Supplementation, and Risk of Hypospadias: Case-Control Study," *Environmental Health Perspectives* (2009), vol. 117, no. 2.

100 *more than five hundred pregnant women fanned out across the South Bronx*: Frederica P. Perera et al., "Effects of Transplacental Exposure to Environ-

mental Pollutants on Birth Outcomes in a Multiethnic Population," *Environmental Health Perspectives* (2003), vol. 111, no. 2.

102 *three times the incidence of heart malformations and valve defects*: Beate Ritz and others, "Ambient Air Pollution and Risk of Birth Defects in Southern California," *American Journal of Epidemiology* (2002), vol. 155, no. 1.

103 *the streams of affluent suburbs burble with discarded Prozac*: Bob Downing, "Medicines, Household Chemicals Flow into Creeks," *Akron Beacon Journal*, February 6, 2005.

103 *traces of flame retardant turn up in the bodies of polar bears*: Derek C. Muir, "Brominated Flame Retardants in Polar Bears (*Ursus maritimus*) from Alaska, the Canadian Arctic, East Greenland, and Svalbard," *Environmental Science and Technology* (2006), vol. 40, no. 2.

103 *the umbilical cords of newborn babies contain an average of 200 industrial chemicals*: Environmental Working Group, "Body Burden: The Pollution in Newborns," July 14, 2005, available at http://reports.ewg.org/reports/bodyburden2/.

104 *some 80 percent of major chemicals used in industry*: Children's Environmental Health Center, "Environmental Toxins," on the website of the Mount Sinai Medical Center, available at www.mountsinai.org/patient-care/service-areas/children/areas-of-care/childrens-environmental-health-center/environmental-toxins.

105 *raising alarms about a "drug drought" in maternal medicine*: Nicholas M. Fisk and Rifat Atun, "Market Failure and the Poverty of New Drugs in Maternal Health," *PLoS Medicine* (2008), vol. 5, no. 1.

107 *"Many of these women were terrified to have another baby"*: Author interview with Nicholas Wald.

107 *women who took the folic acid were 72 percent less likely*: MRC Vitamin Study Research Group, "Prevention of Neural Tube Defects: Results of the Medical Research Council Vitamin Study," *Lancet* (1991), vol. 338, no. 8760.

FIVE MONTHS

110 *a proper pregnancy produced a boy while a girl was evidence of error*: See, e.g., Maryanne Cline Horowitz, "Aristotle and Woman," *Journal of the History of Biology* (1976), vol. 9, no. 2.

110 *John Arbuthnot was physician to the Queen of England*: See, e.g., Russell Bruce Campbell, "John Graunt, John Arbuthnot, and the Human Sex Ratio," *Human Biology* (2001), vol. 72, no. 4, and Daniel R. Headrick, *When Information Came of Age: Technologies of Knowledge in the Age of Reason* (New York: Oxford University Press, 2000), p. 64.

110 *the ratio of male babies to female babies was remarkably consistent*: John Arbuthnot, "An Argument for Divine Providence, taken from the Constant Regularity observed in the Births of both Sexes," *Philosophical Transactions* (1710), vol. 27.

111 *a scientific landmark, the first known use of inferential statistics*: See, e.g., Campbell, "John Graunt, John Arbuthnot, and the Human Sex Ratio," and Chris Pritchard, "The Contributions of Four Scots to the Early Development of Statistics," *The Mathematical Gazette* (1992), vol. 76, no. 475.

111 *when Egyptologists translated the 3,000-year-old scroll*: See, e.g., Lars L. Cederqvist and Fritz Fuchs, "Antenatal Sex Determination: A Historical Review," *Clinical and Obstetric Gynecology* (1970), vol. 13, no. 1, and Thomas R. Forbes, "The Prediction of Sex: Folklore and Science," *Proceedings of the American Philosophical Society* (1959), vol. 103, no. 4.

111 *a twentieth-century re-do*: Office of NIH History, "A Timeline of Pregnancy Testing," http://history.nih.gov/exhibits/thinblueline/time line.html.

112 *alleged indicators of fetal sex, gathered from my reading*: See, e.g., Cederqvist and Fuchs, "Antenatal Sex Determination: A Historical Review"; Forbes, "The Prediction of Sex: Folklore and Science"; and Emmanouil Galanakis, "Sickness and Sex of Child," *Lancet* (2000), vol. 355, no. 9205.

112 *Rarely have the stakes been higher than they were for Anne Boleyn*: See, e.g., Joanna Denny, *Anne Boleyn: A New Life of England's Tragic Queen* (Cambridge, MA: Da Capo Press, 2007); Eric Ives, *The Life and Death of Anne Boleyn* (Malden, MA: Blackwell Publishing, 2005); and Retha W. Warnicke, *The Rise and Fall of Anne Boleyn: Family Politics at the Court of Henry VIII* (Cambridge: Cambridge University Press, 1991).

113 *"quickened at my Lord Gerard's at dinner, and cried out that she was undone"*: Samuel Pepys, *Diary and Correspondence of Samuel Pepys*, Vol. 3 (New York: Dodd and Mead, 1884), p. 72.

114 *"begins in the contemplation of law as soon as an infant is able to stir in the mother's womb"*: William Blackstone, *Commentaries on the Laws of England* (Birmingham, AL: Legal Classics Library, 1983), p. 125.

114 *abortion was not regarded as a crime if it occurred before quickening*: See, e.g., James C. Mohr, *Abortion in America: The Origins and Evolution of National Policy* (New York: Oxford University Press, 1979).

114 *several centuries of artists' imaginings of the fetus*: See, e.g., Alice Elaine Adams, *Reproducing the Womb: Images of Childbirth in Science, Feminist Theory, and Literature* (Ithaca: Cornell University Press, 1994); Norma Broude and Mary D. Garrard, *The Expanding Discourse: Feminism and Art History* (Boulder, CO: Westview Press, 1992); and Karen Newman, *Fetal*

Positions: Individualism, Science, Visuality (Stanford, CA: Stanford University Press, 1996).

115 *"it was not bettered for more than two centuries"*: Peter M. Dunn, "Perinatal Lessons from the Past: Leonardo da Vinci and Reproductive Anatomy," *Archives of Disease in Childhood* (1997), vol. 77, no. 3.

115 *physicians like the German doctor Wilhelm Gottfried Ploucquet*: Barbara Duden, "Quick with Child: An Experience That Has Lost Its Status," *Technology in Society* (1992), vol. 14, no. 3.

116 *French doctor Jacques Alexandre Le Jumeau de Kergaradec*: See, e.g., Donald Gibb and Sabaratnam Arulkumaran, *Fetal Monitoring in Practice* (New York: Elsevier, 1997), p. 1, and Michael J. O'Dowd and Elliot Elias Philipp, *The History of Obstetrics and Gynecology* (New York: Parthenon, 1994), p. 97.

116 *the heart rate of the fetus became one more way to guess its sex*: See, e.g., Cederqvist and Fuchs, "Antenatal Sex Determination: A Historical Review."

116 *growing understanding of the genetic basis of sex-linked diseases such as hemophilia*: See, e.g., Marion S. Verp and Joe Leigh Simpson, "Amniocentesis for Prenatal Genetic Diagnosis," in *Human Prenatal Diagnosis*, edited by Karen Filkins and Joseph F. Russo (New York: M. Dekker, 1990).

116 *the nuclei of cells from female humans contained distinctive structures called sex chromatins*: Murray L. Barr and Ewart G. Bertram, "A Morphological Distinction Between Neurones of the Male and Female, and the Behaviour of the Nuclear Satellites During Accelerated Nucleoprotein Synthesis," *Nature* (1949), vol. 163.

117 *in one of those striking feats of scientific convergence*: See, e.g., Fritz Fuchs and Povl Riis, "Antenatal Sex Determination," *Nature* (1956), vol. 177, no. 4503, and Povl Riis, "First Steps in Antenatal Diagnosis, 1956," *Human Genetics* (2006), vol. 118, no. 6.

117 *a young doctor named Ian Donald*: See, e.g., Asim Kurjack, "Ultrasound Scanning—Prof. Ian Donald (1910–1987)," *European Journal of Obstetrics and Gynecology and Reproductive Biology* (2000), vol. 90; Margaret B. McNay and John E.E. Fleming, "Forty Years of Obstetric Ultrasound 1957–1997: From A-Scope to Three Dimensions," *Ultrasound in Medicine and Biology* (1999), vol. 25, no. 1; Alistair G.S. Philip, "Ian Donald and Obstetric Diagnostic Ultrasound," *NeoReviews* (2007), vol. 8, no. 5; and Edward Yoxen, "Seeing with Sound: A Study of the Development of Medical Images," in *The Social Construction of Technological Systems*, edited by Wiebe Bijker, Thomas P. Hughes, and Trevor Pinch (Cambridge, MA: MIT Press, 1987).

117 *"mad, red-headed Scotsman"*: Philip, "Ian Donald and Obstetric Diagnostic Ultrasound."

NOTES

117 *"I've always been an engineer at heart"*: Quoted in Ann Oakley, *The Captured Womb: A History of the Medical Care of Pregnant Women* (New York: Blackwell Publishers, 1985), p. 157.

117 *"There is not so much difference, after all, between a fetus in utero and a submarine at sea"*: Ian Donald, "On Launching a New Diagnostic Science," *American Journal of Obstetrics and Gynecology* (1969), vol. 103, no. 5.

118 *"such a situation, in the patient's own bed, had all the ingredients of a shambles"*: Ian Donald, "Sonar—The Story of an Experiment," *Ultrasound Medical Biology* (1974), vol. 1, no. 2.

118 *"said he did not know but would run out to the car"*: Donald, "Sonar—The Story of an Experiment."

118 *"I will not forget the excitement with which I observed the first very early pregnancy"*: Donald, "Sonar—The Story of an Experiment."

118 *"the developing fetus has been hidden behind a veritable iron curtain"*: Ian Donald, "Sonar as a Method of Studying Prenatal Development," *Journal of Pediatrics* (1969), vol. 75, no. 2.

118 *The pictures he produced were shown to the world for the first time*: Ian Donald, John Macvicar, and Tom Brown, "Investigation of Abdominal Masses by Pulsed Ultrasound," *Lancet* (1958), vol. 1, no. 7032.

119 *one rival in Edinburgh sniped*: Malcolm Nicolson, "Ian Donald—Diagnostician and Moralist," Goodall Memorial Lecture, given to the Royal College of Physicians and Surgeons of Glasgow, June 5, 2000.

119 *"The first forty weeks of existence may well prove to be far more important medically than the next forty years"*: Ian Donald, "Sonar as a Method of Studying Prenatal Development," *Journal of Pediatrics* (1969), vol. 75, no. 2.

120 *"technological quickening"*: Duden, "Quick with Child: An Experience That Has Lost its Status." See also Barbara Duden, *Disembodying Women: Perspectives on Pregnancy and the Unborn* (Cambridge, MA: Harvard University Press, 1993).

120 *"I was inspired by the pediatrician T. Berry Brazelton"*: Author interview with Zack Boukydis.

120 *tested the effects of the enhanced ultrasound on pregnant women's psychological state*: Zach Boukydis, "Ultrasound Consultation to Reduce Risk and Increase Resilience in Pregnancy," *Annals of the New York Academy of Sciences* (2006), vol. 1094.

121 *the role of ultrasound in helping expectant mothers "bond" with their fetuses*: See, e.g., Judith Lumley, "Through a Glass Darkly: Ultrasound and Prenatal Bonding," *Birth* (1990), vol. 17, no. 4, and Stuart Campbell, "4D and Prenatal Bonding: Still More Questions Than Answers," *Ultrasound in Obstetrics and Gynecology* (2006), vol. 27, no. 3.

121 *When the notion of maternal bonding was first introduced*: Marshall H. Klaus

and others, "Maternal Attachment: Importance of the First Postpartum Days," *New England Journal of Medicine* (1972), vol. 286. See also Janelle S. Taylor, "Image of Contradiction: Obstetrical Ultrasound in American Culture," in *Reproducing Reproduction: Kinship, Power, and Technological Innovation*, edited by Sarah Franklin and Helena Ragoné (Philadelphia: University of Pennsylvania Press, 1998).

122 *an old issue of* Life *magazine*: Lennart Nilsson, "Drama of Life Before Birth," *Life Magazine*, April 30, 1965.

122 *he made the pregnant woman disappear*: See, e.g., Barbara Duden, *Disembodying Women: Perspectives on Pregnancy and the Unborn*, and Nathan Stormer, "Seeing the Fetus: The Role of Technology and Image in the Maternal-Fetal Relationship," *Journal of the American Medical Association* (2003), vol. 289, no. 13.

123 *"Manipulation was crucial in the construction of the famous cover image"*: Clare Hanson, *A Cultural History of Pregnancy: Pregnancy, Medicine and Culture, 1750–2000* (New York: Palgrave Macmillan, 2004), p. 156.

123 *Their efforts fill a stack of books on my desk*: Peter Doubilet, Carol Benson, and Roanne Weisman, *Your Developing Baby, Conception to Birth: Witnessing the Miraculous 9-Month Journey* (New York: McGraw-Hill, 2008); Lennart Nilsson, *A Child Is Born* (New York: Delta, 2004); Peter Tallack, *In the Womb: Witness the Journey from Conception to Birth Through Astonishing 3D Images* (New York: National Geographic, 2006); and Alexander Tsiaras, *From Conception to Birth: A Life Unfolds* (New York: Doubleday, 2002).

124 *I slide a series of discs into my DVD player*: "From Conception to Birth," produced by the Discovery Channel, 2007; "In the Womb," produced by National Geographic, 2005; and "The Miracle of Life," produced by NOVA, 2000.

124 *"Doctors have discovered that for parents-to-be, the scans play a beneficial role"*: National Geographic, "In the Womb."

125 *as many as 80 percent of women choose to find out the sex of their babies*: See, e.g., Kevin Harrington et al., "Fetal Sexing by Ultrasound in the Second Trimester: Maternal Preference and Professional Ability," *Ultrasound in Obstetrics and Gynecology* (1996), vol. 8, no. 5, and Deborah F. Perry, Janet DiPietro, and Kathleen Costigan, "Are Women Carrying 'Basketballs' Really Having Boys?: Testing Pregnancy Folklore," *Birth: Issues in Perinatal Care* (1999), vol. 26, no. 3.

125 *Women who are pregnant with girls may indeed experience more nausea*: See, e.g., John Askling et al., "Sickness in Pregnancy and Sex of Child," *Lancet* (1999), vol. 354, no. 9195, and Maria del Mar Melero-Montes and Hershel Jick, "Hyperemesis Gravidarum and the Sex of the Offspring," *Epidemiology* (2000), vol. 12, no. 1.

125 *"can say with confidence to a woman who has hyperemesis gravidarum"*: A.A. Verkuyl Douwe, "Sickness and Sex of Child," *Lancet* (2000), vol. 355, no. 9205.

125 *the sicker she is, the more likely this is to be true*: Melissa A. Schiff, Susan D. Reed, and Janet R. Daling, "The Sex Ratio of Pregnancies Complicated by Hospitalisation for Hyperemesis Gravidarum," *British Journal of Obstetrics and Gynecology* (2004), vol. 111, no. 1.

126 *the diets of more than 300 women receiving prenatal care*: Rulla M. Tamimi et al., "Average Energy Intake Among Pregnant Women Carrying a Boy Compared with a Girl," *British Medical Journal* (2003), vol. 326, no. 7401.

126 *"contrary to expectations," women who rely on dreams and emotions to guess their babies' sex*: Perry, DiPietro, and Costigan, "Are Women Carrying 'Basketballs' Really Having Boys?: Testing Pregnancy Folklore."

127 *only an estimated 20 to 40 percent of fertilized eggs actually results in a live birth*: John M. Opitz, "Early Embryonic Development: An Up-to-Date Account," testimony given to the Presidential Commission for the Study of Bioethical issues, January 16, 2003, available at http://www11.george town.edu/research/nrcbl/pcbe/transcripts/jan03/session1.html.

127 *a "killer fog" descended on much of London*: See, e.g., Michelle L. Bell, Devra L. Davis, and Tony Fletcher, "A Retrospective Assessment of Mortality from the London Smog Episode of 1952: The Role of Influenza and Pollution," *Environmental Health Perspectives* (2004), vol. 112, no. 1, and W.R. Lyster, "Altered Sex Ratio After the London Smog of 1952 and the Brisbane Flood of 1965," *Journal of Obstetrics and Gynaecology of the British Commonwealth* (1974), vol. 81, no. 8.

128 *It happened after the East German economy collapsed*: Ralph A. Catalano, "Sex Ratios in the Two Germanies: A Test of the Economic Stress Hypothesis," *Human Reproduction* (2003), vol. 13, no. 9.

128 *It happened after a devastating earthquake struck Kobe, Japan*: Misao Fukuda and others, "Decline in Sex Ratio at Birth After Kobe Earthquake," *Human Reproduction* (1998), vol. 13, no. 8.

128 *It may even have happened after terrorists flew planes into the World Trade Center towers*: Ralph Catalano and others, "Exogenous Shocks to the Human Sex Ratio: The Case of September 11, 2001, in New York City," *Human Reproduction* (2006), vol. 21, no. 12.

128 *"From an evolutionary perspective, in stressful eras it's a smarter bet to have female child"*: Author interview with Ralph Catalano.

129 *a study he made of the life spans of Swedish men born between 1751 and 1912*: Ralph Catalano and Tim Bruckner, "Secondary Sex Ratios and Male Lifespan: Damaged or Culled Cohorts," *Proceedings of the National Academies of Science* (2006), vol. 103, no. 5.

129 *members of the tribe began voicing concern that fewer and fewer boys were being born*: See, e.g., Judith Graham, "A Puzzle over Fewer Boy Births," *Chicago Tribune*, May 25, 2008, and Martin Mittelstaedt, "Humanity at Risk: Are the Males Going First?" *Globe and Mail*, September 20, 2008.

129 *male births among the Aamjiwnaang had begun falling in the early 1990s*: Constanze A. Mackenzie, Ada Lockridge, and Margaret Keith, "Declining Sex Ratio in a First Nation Community," *Environmental Health Perspectives* (2005), vol. 113, no. 10.

130 *Sex ratios in a number of industrialized countries appear to be dropping*: See, e.g., Devra L. Davis, Michell B. Gottlieb, and Julie R. Stampnitzky, "Reduced Ratio of Male to Female Births in Several Industrial Countries: A Sentinel Health Indicator?" *Journal of the American Medical Association* (1998), vol. 279, no. 13.

130 *there were 1,059 male births for every 1,000 female births in 1946*: National Center for Health Statistics, "More Boys Born Than Girls: New Report Documents Total Gender Ratios at Birth From 1940 to 2002," press release, June 14, 2005.

130 *a "sentinel health indicator"*: See, e.g., Davis, Gottlieb, and Stampnitzky, "Reduced Ratio of Male to Female Births in Several Industrial Countries: A Sentinel Health Indicator?"; and Nicolas van Larebeke and others, "Sex Ratio Changes as Sentinel Health Events of Endocrine Disruption," *International Journal of Occupational and Environmental Health* (2008), vol. 14, no. 2.

130 *"We're talking about the reproduction of the species here"*: Author interview with Devra Davis.

130 *the sex ratio declined significantly among white Americans*: Devra Lee Davis et al., "Declines in Sex Ratio at Birth and Fetal Deaths in Japan, and in U.S. Whites but Not African Americans," *Environmental Health Perspectives* (2007), vol. 115, no. 6.

131 *female animals who are stressed have fewer male offspring*: See, e.g., Nancy C. Pratt and Robert D. Lisk, "Effects of Social Stress During Early Pregnancy on Litter Size and Sex Ratio in the Golden Hamster (*Mesocricetus auratus*)," *Journal of Reproduction and Fertility* (1989), vol. 87, no. 2, and Loeske E. B. Kruuk et al., "Population Density Affects Sex Ratio Variation in Red Deer," *Nature* (1999), vol. 399, no. 6735.

131 *six thousand Danish women evaluated for psychological distress in early pregnancy*: Carsten Obel et al., "Psychological Distress During Early Gestation and Offspring Sex Ratio," *Human Reproduction* (2007), vol. 22, no. 11.

131 *the more antidepressants are prescribed to women . . . the fewer boys are born*: Ralph Catalano et al., "Population Stress and the Swedish Sex Ratio," *Paediatric and Perinatal Epidemiology* (2005), vol. 19, no. 6.

131   *women who are not living with a man at the time of conception*: Karen Norberg, "Partnership Status and the Human Sex Ratio at Birth," *Proceedings of the Royal Society B: Biological Sciences* (2004), vol. 271, no. 1555.

131   *women's nutrition around the time of conception*: Fiona Mathews, Paul J. Johnson, and Andrew Neil, "You Are What Your Mother Eats: Evidence for Maternal Preconception Diet Influencing Foetal Sex in Humans," *Proceedings of the Royal Society B: Biological Sciences* (2008), vol. 275, no. 1643.

132   *Mathews's study soon came in for a scathing rebuke*: Stanley S. Young, Heejung Bang, and Kutluk Oktay, "Cereal-Induced Gender Selection? Most Likely a Multiple Testing False Positive," *Proceedings of the Royal Society B: Biological Sciences* (2009), vol. 276, no. 1660.

132   *Mathews defended her hypothesis*: Fiona Mathews, Paul J. Johnson, and Andrew Neil, "Reply to Comment by Young et al.," *Proceedings of the Royal Society B: Biological Sciences* (2009), vol. 276, no. 1660.

132   *humankind has long wished to deliberately choose the sex of children*: See, e.g., Hymie Gordon, "Ancient Ideas About Sex Differentiation," in *Genetic Mechanisms of Sexual Development*, edited by H. Lawrence Vallet and Ian H. Porter (New York: Academic Press, 1979), and Eugene S. McCartney, "Sex Determination and Sex Control in Antiquity," *American Journal of Philology* (1922), vol. 43.

133   *In today's America, it most likely means reading Dr. Shettles*: See, e.g., Robin Marantz Henig, "Second Best: Landrum B. Shettles," *New York Times Magazine*, December 28, 2003, and Stuart Lavietes, "Dr. L. B. Shettles, 93, Pioneer in Human Fertility," *New York Times*, February 16, 2003.

133   *Shettles was among the researchers*: Landrum B. Shettles, "Nuclear Morphology of Cells in Human Amniotic Fluid in Relation to Sex of Infant," *American Journal of Obstetrics and Gynecology* (1956), vol. 71, no. 4.

133   *"They imagined he might be sympathetic to their longings"*: Landrum Brewer Shettles and David M. Rorvik, *How to Choose the Sex of Your Baby: The Method Best Supported by Scientific Evidence* (New York: Random House, 1996), p. 7.

134   *a success rate as high as 85 percent*: Shettles and Rorvik, *How to Choose the Sex of Your Baby*, p. 164.

134   *worried that Shettles was making a "laughingstock" of the department*: Robin Marantz Henig, *Pandora's Baby: How the First Test-Tube Babies Sparked the Reproductive Revolution* (New York: Houghton Mifflin Harcourt, 2004), p. 60.

134   *"the timing of sexual intercourse in relation to ovulation has no influence on the sex of the baby"*: Allen J. Wilcox et al., "Timing of Sexual Intercourse in Relation to Ovulation—Effects on the Probability of Conception, Survival of the Pregnancy, and Sex of the Baby," *New England Journal of Medicine* (1995), vol. 333, no. 23.

134 *"There's zero evidence that the Shettles method works"*: Author interview with Allen Wilcox.

134 *eight percent of the 243 pregnant women surveyed had tried to influence the sex of their baby*: Molly Kay Walker and Garris Keels Conner, "Fetal Sex Preference of Second-Trimester Gravidas," *Journal of Nurse Midwifery* (1993), vol. 38, no. 2.

134 *The Shettles method was once put to a real-life test*: Nancy E. Williamson, T.H. Lean, and D. Vengadasalam, "Evaluation of an Unsuccessful Sex Preselection Clinic in Singapore," *Journal of Biosocial Science* (1978), vol. 10.

135 *the number of female fetuses aborted over the past twenty years*: See, e.g., Sabu M. George, "Millions of Missing Girls: From Fetal Sexing to High Technology Sex Selection in India," *Prenatal Diagnosis* (2006), vol. 26, no. 7, and Theresa M. Marteau and Lyn S. Chitty, "Sex Selection: Triumph or Tyranny?" *Prenatal Diagnosis* (2006), vol. 26, no. 7.

135 *these rootless young men may turn to violence, militarism, or even terrorism*: See, e.g., Therese Hesket and Zhu Wei Xing, "Abnormal Sex Ratios in Human Populations: Causes and Consequences," *Proceedings of the National Academies of Science* (2006), vol. 103, no. 36, and Valerie M. Hudson and Andrea M. den Boer, *Bare Branches: The Security Implications of Asia's Surplus Male Population* (Cambridge, MA: MIT Press, 2004).

135 *evidence that sex-selective abortion goes on in the U.S.*: Douglas Almond and Lena Edlund, "Son-Biased Sex Ratios in the 2000 United States Census," *Proceedings of the National Academy of Sciences* (2008), vol. 105, no. 15. See also Jason Abrevaya, "Are There Missing Girls in the United States?: Evidence from Birth Data," *American Economic Journal: Applied Economics* (2009), vol. 1, no. 2.

136 *Forty percent of U.S. fertility clinics already offer this option to their clients*: Susannah Baruch, David Kaufman, and Kathy L. Hudson, "Genetic Testing of Embryos: Practices and Perspectives of U.S. In Vitro Fertilization Clinics," *Fertility and Sterility* (2008), vol. 89, no. 5.

136 *almost 10 percent of preimplantation genetic diagnoses . . . are performed for gender selection alone*: Baruch, Kaufman, and Hudson, "Genetic Testing of Embryos: Practices and Perspectives of U.S. In Vitro Fertilization Clinics."

136 *respondents were asked if they would choose the sex of a child*: Edgar Dahl et al., "Preconception Sex Selection Demand and Preferences in the United States," *Fertility and Sterility* (2006), vol. 85, no. 2.

137 *describe the movements of their fetuses*: Barbara Katz Rothman, *The Tentative Pregnancy: Prenatal Diagnosis and the Future of Motherhood* (New York: Viking, 1986), p. 129.

137 *"I used to say, along with most social scientists, that gender socialization begins at birth"*: Rothman, *The Tentative Pregnancy*, p. 128.

138 *high-speed 3D ultrasound offers the clearest picture yet*: See, e.g., Michelle Jean-
dron, "Making Waves: The 3D Ultrasound Pioneer," *Medical Physics Web*,
October 10, 2007.

138 *A team of Japanese scientists, led by Toshiyuki Hata*: See, e.g., Toshiyuki Hata
et al., "Real-Time 3-D Sonographic Observation of Fetal Facial Expression," *Journal of Obstetrics and Gynaecology Research* (2005), vol. 31, no.
4, and Fang Yan et al., "Four-Dimensional Sonographic Assessment of
Fetal Facial Expression Early in the Third Trimester," *International Journal of Gynaecology and Obstetrics* (2006), vol. 94, no. 2.

138 *"may be a key to predicting fetal brain function and well-being"*: Yan, "Four-
Dimensional Sonographic Assessment of Fetal Facial Expression Early
in the Third Trimester."

SIX MONTHS

142 *"a guerrilla war with my own life"*: Joan Didion, *The White Album* (New
York: Macmillan, 1990), p. 172.

143 *researchers have proposed that the* Mona Lisa . . . *was pregnant*: See, e.g., Jean-
Pierre Mohen, Michel Menu, and Bruno Mottin, *Mona Lisa: Inside the
Painting* (New York: Harry N. Abrams, 2006).

143 *"inner satisfaction that the miracle of life is being created within her body"*: Sherwin Nuland, *Leonardo da Vinci* (New York: Viking, 2000), p. 80.

143 *"a smile of the coming of life"*: Sherwin Nuland, quoted in Richard
Brooks, "Mona Lisa Had an Air of Expectation," *Times* (London), April
6, 2003.

144 *"are insane on every pregnancy or lying-in, others only occasionally"*: George
Man Burrows, *Commentaries on the Causes, Forms, Symptoms, and Treatment,
Moral and Medical, of Insanity* (London: Underwood, 1828), p. 364.

144 *"The marked perturbation of the nervous system"*: William Thompson Lusk,
The Science and Art of Midwifery (New York: D. Appleton, 1892), p. 701.

145 *"how to have handsome children"*: Quoted in Philip K. Wilson, "Eighteenth-
Century 'Monsters' and Nineteenth-Century 'Freaks': Reading the
Maternally Marked Child," in *Literature and Medicine* (2002), vol. 21,
no. 1.

145 *"It is time for me to grow more reasonable"*: Mary Wollstonecraft, *Letters to
Imlay* (London: C. Kegan Paul, 1879), p. 27.

145 *the tale of Mary Toft*: See, e.g., Bill Bynum, "Maternal Impressions,"
Lancet (2002), vol. 359, no. 898, and T.E.C. Jr., "Mary Toft of Godly-
man, England, and Her Extraordinary Delivery of Seventeen Rabbits in
1726," *Pediatrics* (1980), vol. 66, no. 4.

146 *the pregnant wife of the sixteenth-century German botanist Joachim Camerarius*: Cited in Julia Epstein, "The Pregnant Imagination, Fetal Rights, and Women's Bodies: A Historical Inquiry," *Yale Journal of Law and the Humanities* (1995), vol. 7, no. 139.

147 *Blondel and Turner were soon embroiled in a "pamphlet war"*: See, e.g., Philip K. Wilson, "'Out of Sight, Out of Mind?': The Daniel Turner–James Blondel Dispute Over the Power of the Maternal Imagination," *Annals of Science* (1992), vol. 49, no. 1.

148 *pregnant women who experienced morning sickness*: See, e.g., William A. Harvey and Mary Jane Sherfey, "Vomiting in Pregnancy: A Psychiatric Study," *Psychosomatic Medicine* (1954), vol. 16, no. 1, and Sidney Rosen, "Emotional Factors in Nausea and Vomiting of Pregnancy," *Psychiatric Quarterly* (1955), vol. 29, no. 4.

149 *the case of a patient she called "Mrs. Smith"*: Helene Deutsch, *The Psychology of Women: A Psychoanalytic Interpretation*, Vol. 2 (New York: Grune and Stratton, 1945), pp. 155–156.

149 *Mrs. Smith, Deutsch confessed years later, was really the analyst herself*: Helene Deutsch, *Confrontations with Myself: An Epilogue* (New York: Norton, 1973). See also Paul Roazen, *Helene Deutsch: A Psychoanalyst's Life* (New York: Anchor Press, 1985).

150 *"For a number of years an interest has been growing increasingly"*: Grete L. Bibring, "Some Considerations of the Psychological Processes in Pregnancy," *The Psychoanalytic Study of the Child* (1959), vol. 14.

150 *"the folklore and old wives' tales"*: Grete L. Bibring and Arthur F. Valenstein, "Psychological Aspects of Pregnancy," *Clinical Obstetrics and Gynecology* (1976), vol. 19, no. 2.

150 *a "normal crisis," a "point of no return between one phase and the next"*: Bibring, "Some Considerations of the Psychological Processes in Pregnancy."

151 *"There is no turning back" from this developmental challenge, Bibring concludes; pregnancy is a "testing ground of psychological health"*: Bibring and Valenstein, "Psychological Aspects of Pregnancy."

151 *from my reading that exercise can alleviate depression during pregnancy*: See, e.g., Deborah Da Costa et al., "Self-Reported Leisure-Time Physical Activity During Pregnancy and Relationship to Psychological Well Being," *Journal of Psychosomatic Obstetrics and Gynecology* (2003), vol. 24, no. 2; Danielle Symons Downs, Jennifer M. DiNallo, and Tiffany L. Kirner, "Determinants of Pregnancy and Postpartum Depression: Prospective Influences of Depressive Symptoms, Body Image Satisfaction, and Exercise Behavior," *Annals of Behavioral Medicine* (2008), vol. 36, no. 1; and Mélanie S. Poudevigne and Patrick J. O'Connor, "A Review of Physical

Activity Patterns in Pregnant Women and Their Relationship to Psychological Health," *Sports Medicine* (2006), vol. 35, no. 1.

151 *Research suggests that it may also reduce the risk of preeclampsia and gestational diabetes, and help manage musculoskeletal problems like low back pain*: See, e.g., The American College of Sports Medicine, "Impact of Physical Activity During Pregnancy and Postpartum on Chronic Disease Risk," *Medicine and Science in Sports and Exercise* (2006), vol. 38, no. 5.

152 *May conducted a study of the effects of maternal exercise on the fetus*: Linda May et al., "Effects of Maternal Exercise on the Fetal Heart," *The FASEB Journal* (2008), vol. 1175, no. 3.

152 *One of his principal pieces of advice for raising smarter kids: work out while you're pregnant*: Richard Nisbett, *Intelligence and How to Get It: Why Schools and Cultures Count* (New York: W. W. Norton, 2009), p. 184.

153 *mood disorders are no more common among pregnant women than among their non-pregnant peers. But it's during a woman's childbearing years—roughly 20 to 40—that she is at the highest risk of depression or anxiety*: Oriana Vesga-López et al., "Psychiatric Disorders in Pregnant and Postpartum Women in the United States," *Archives of General Psychiatry* (2008), vol. 65, no. 7: ". . . although rates of Axis I psychiatric disorders, including substance use, mood, and anxiety disorders, are high in women of childbearing age regardless of pregnancy status, pregnancy per se is not associated with an increased risk of new onset or recurrence of the most prevalent mental disorders."

153 *contrary to what doctors once believed, pregnancy confers no protection*: Shari I. Lusskin, Tara M. Pundiak, and Sally M. Habib, "Perinatal Depression: Hiding in Plain Sight," *Canadian Journal of Psychiatry* (2007), vol. 52, no. 8.

152 *about 20 percent of pregnant women experience mood or anxiety disorders and about 10 percent develop major depression*: The Mood and Anxiety Disorders Resource Center, Massachusetts General Hospital, available at http://www2.massgeneral.org/madiresourcecenter/moodandanxiety_considering-treatment-options_pregnancyandchildbirth.asp.

153 *"More mothers moved above the threshold for depression between eighteen weeks and thirty-two weeks of pregnancy"*: Jonathan Evans et al., "Cohort Study of Depressed Mood During Pregnancy and After Childbirth," *British Medical Journal* (2001), vol. 323, no. 7307.

153 *About half of women depressed during pregnancy will be depressed after birth as well*: Patricia M. Dietz and others, "Clinically Identified Maternal Depression Before, During, and After Pregnancies Ending in Live Births," *American Journal of Psychiatry* (2007), vol. 164, no. 10.

153 *"the single most predictive factor for postpartum depression"*: Amy Salisbury and others, "Maternal-Fetal Psychobiology: A Very Early Look at Emotional Development," in *Emotional Development: Recent Research Advances*, edited by Jacqueline Nadel and Darwin Muir (New York: Oxford University Press, 2005), p. 97.

154 *the use of SSRIs during pregnancy has been associated in some studies with an increase in breathing problems in the newborn*: See, e.g., Tim F. Oberlander et al., "Neonatal Outcomes After Prenatal Exposure to Selective Serotonin Reuptake Inhibitor Antidepressants and Maternal Depression Using Population-Based Linked Health Data," *Archives of General Psychiatry* (2006), vol. 63, no. 8, and Tim F. Oberlander et al., "Effects of Timing and Duration of Gestational Exposure to Serotonin Reuptake Inhibitor Antidepressants: Population-Based Study," *British Journal of Psychiatry* (2008), vol. 192, no. 5.

154 *"Pregnant depressed women are really on the horns of a dilemma"*: Author interview with Shari Lusskin.

154 *followed 201 pregnant women with a history of major depression*: Lee S. Cohen et al., "Relapse of Major Depression During Pregnancy in Women Who Maintain or Discontinue Antidepressant Treatment," *Journal of the American Medical Association* (2006), vol. 295, no. 5.

154 *women with mild symptoms of depression are 60 percent more likely to deliver early; women who are severely depressed have twice the risk of premature birth*: De-Kun Li, Liyan Liu, and Roxana Odouli, "Presence of Depressive Symptoms During Early Pregnancy and the Risk of Preterm Delivery: A Prospective Cohort Study," *Human Reproduction* (2009), vol. 24, no. 1.

155 *"New research indicates that even before birth, women's moods may affect child development"*: Catherine Monk, "The Putative Role of Cortisol in the 'Transmission' of Maternal Stress/Anxiety/Depression," presentation given at the International Perinatal Brain and Behavior Network Symposium, November 12, 2008.

158 *the fetuses of depressed and anxious women are especially reactive*: See, e.g., Catherine Monk et al., "Effects of Women's Stress-Elicited Physiological Activity and Chronic Anxiety on Fetal Heart Rate," *Journal of Developmental and Behavioral Pediatrics* (2003), vol. 24, no. 1.

159 *the newborns of depressed mothers are more irritable and hard to soothe*: Elysia Poggi Davis and others, "Prenatal Exposure to Maternal Depression and Cortisol Influences Infant Temperament," *Journal of the American Academy of Child and Adolescent Psychiatry* (2007), vol. 46, no. 6.

159 *the newborns of depressed mothers . . . have more problems sleeping*: See, e.g., Roseanne Armitage et al., "Early Developmental Changes in Sleep in Infants: The Impact of Maternal Depression," *Sleep* (2009), vol. 32, no. 5.

159   *the newborns of depressed mothers . . . have higher levels of the stress hormone cortisol in their blood*: See, e.g., Miguel A. Diego et al., "Prepartum, Postpartum, and Chronic Depression Effects on Newborns," *Psychiatry* (2004), vol. 67, no. 1, and Brenda L. Lundy et al., "Prenatal Depression Effects on Neonates," *Infant Behavior and Development* (1999), vol. 22, no. 1.

159   *maternal depression and anxiety during pregnancy is associated with higher rates of impulsivity, hyperactivity, and emotional and behavioral problems*: See, e.g., Toity Deave et al., "The Impact of Maternal Depression in Pregnancy on Early Child Development," *British Journal of Obstetrics and Gynecology* (2008), vol. 115, no. 8; Thomas G. O'Connor et al., "Antenatal Anxiety Predicts Child Behavioral/Emotional Problems Independently of Postnatal Depression," *Journal of the American Academy of Child and Adolescent Psychiatry* (2002), vol. 41, no. 12; Kieran O'Donnell, Thomas G. O'Connor, and Vivette Glover, "Prenatal Stress and Neurodevelopment of the Child: Focus on the HPA Axis and Role of the Placenta," *Developmental Neuroscience* (2009), vol. 31, no. 4; and Bea R. H. Van den Bergh et al., "Antenatal Maternal Anxiety and Stress and the Neurobehavioural Development of the Fetus and Child: Links and Possible Mechanisms: A Review," *Neuroscience and Biobehavioral Reviews* (2005), vol. 29, no. 2.

161   *examined thirty years of psychiatric case records from the Wuhu region of Anhui*: David St. Clair et al., "Rates of Adult Schizophrenia Following Prenatal Exposure to the Chinese Famine of 1959–1961," *Journal of the American Medical Association* (2005), vol. 294, no. 5.

162   *Individuals in utero at the time of the Nazis' siege of the Netherlands had a twofold risk of developing schizophrenia*: Ezra Susser, "Schizophrenia After Prenatal Famine: Further Evidence," *Archives of General Psychiatry* (1996), vol. 53, no. 1.

162   *Mervyn Susser and Zena Stein, pioneering epidemiologists*: See, e.g., Zena Stein, *Famine and Human Development* (New York: Oxford University Press, 1975).

163   *There are now several instruments available for screening pregnant women for depression*: See, e.g., the Edinburgh Postnatal Depression Scale, the Pregnancy Depression Scale, the Pregnancy Risk Assessment Monitoring System Questionnaire, and the Pregnancy Risk Questionnaire.

163   *The Pregnancy Depression Scale*: Lori L. Altshuler et al., "The Pregnancy Depression Scale (PDS): A Screening Tool for Depression in Pregnancy," *Archives of Women's Mental Health* (2008), vol. 11, no. 4.

164   *"Interpersonal therapy focuses on individuals' relationships"*: Author interview with Margaret Spinelli.

164 *the first controlled clinical treatment trial for therapy during pregnancy*: Margaret G. Spinelli and Jean Endicott, "Controlled Clinical Trial of Interpersonal Psychotherapy Versus Parenting Education Program for Depressed Pregnant Women," *American Journal of Psychiatry* (2003), vol. 160, no. 3.

SEVEN MONTHS

171 *when Barker introduced his idea the reaction was scathing*: Author interview with David Barker. See also David Barker, *The Best Start in Life* (London: Century, 2003), p. 31, and Stephen S. Hall, "Small and Thin: The Controversy over the Fetal Origins of Adult Health," *The New Yorker*, November 19, 2007.

172 *FOAD and DOHaD*: For more, see Peter D. Gluckman and Mark A. Hanson, "The Developmental Origins of Health and Disease: An Overview," in *Developmental Origins of Health and Disease*, edited by Peter D. Gluckman and Mark A. Hanson (Cambridge: Cambridge University Press, 2006).

174 *Goddard published a detailed account of the clan*: Henry Herbert Goddard, *The Kallikak Family: A Study in the Heredity of Feeble-Mindedness* (New York: Macmillan, 1912). See also Leila Zenderland, *Measuring Minds: Henry Herbert Goddard and the Origins of American Intelligence Testing* (Cambridge: Cambridge University Press, 2001).

174 *the book was popular reading among members of the Nazi party in Germany*: Stefan Kuhl, *The Nazi Connection: Eugenics, American Racism, and German National Socialism* (New York: Oxford University Press, 2002), p. 41.

175 *Karp's analysis of the records and photographs he found in the Vineland files*: Robert J. Karp and others, "Fetal Alcohol Syndrome at the Turn of the 20th Century: An Unexpected Explanation of the Kallikak Family," *Archives of Pediatrics and Adolescent Medicine* (1995), vol. 149, no. 1.

175 *"a primal myth of the eugenics movement"*: Stephen Jay Gould, *The Mismeasure of Man* (New York: Norton, 1981), p. 198.

175 *The theories and methods of the eugenicists have long been soundly repudiated*: See, e.g., Scott Christianson, "Bad Seed or Bad Science?: The Story of the Notorious Jukes Family," *New York Times*, February 8, 2003, and Elof Axel Carlson, "R. L. Dugdale and the Jukes Family: A Historical Injustice Corrected," *BioScience* (1980), vol. 30, no. 8.

176 *the stress of childhood poverty can interfere with the development of memory, problem-solving and language skills*: See, e.g., Gary W. Evans and Michelle A. Schamberg, "Childhood Poverty, Chronic Stress, and Adult Working Memory," *Proceedings of the National Academy of Sciences* (2009), vol. 106, no. 13; Daniel A. Hackman and Martha J. Farah, "Socioeconomic Status

and the Developing Brain," *Trends in Cognitive Science* (2009), vol. 13, no. 2; and Mark M. Kishiyama et al., "Socioeconomic Disparities Affect Prefrontal Function in Children," *Journal of Cognitive Neuroscience* (2009), vol. 21, no. 6.

176 *poor children may enter the world already disadvantaged by adversity experienced before birth*: See, e.g., Calvin Hobel and Jennifer Culhane, "Role of Psychosocial and Nutritional Stress on Poor Pregnancy Outcome," *Journal of Nutrition* (2003), vol. 133, no. 5, and Rebekah L. Weck, Tessie Paulose, and Jodi Flaws, "Impact of Environmental Factors and Poverty on Pregnancy Outcomes," *Clinical Obstetrics and Gynecology* (2008), vol. 51, no. 2.

177 *prenatal exposure to nicotine . . . can change the fetal brain in ways that make the offspring more likely to experiment with or become addicted to such drugs as adolescents and adults*: Yael Abreu-Villaça et al., "Prenatal Nicotine Exposure Alters the Response to Nicotine Administration in Adolescence: Effects on Cholinergic Systems During Exposure and Withdrawal," *Neuropsychopharmacology* (2004), vol. 29, no. 5; Abdullah Al Mamun et al., "Does Maternal Smoking During Pregnancy Predict the Smoking Patterns of Young Adult Offspring?: A Birth Cohort Study," *Tobacco Control* (2006), vol. 15, no. 6; Stephen L. Buka, Edmond D. Shenassa, and Raymond Niaura, "Elevated Risk of Tobacco Dependence Among Offspring of Mothers Who Smoked During Pregnancy: A 30-Year Prospective Study," *American Journal of Psychiatry* (2003), vol. 160; Marie D. Cornelius et al., "Prenatal Tobacco Exposure: Is It a Risk Factor for Early Tobacco Experimentation?" *Nicotine and Tobacco Research* (2000), vol. 2, no. 1; and Denise B. Kandel, Ping Wu, and Mark Davies, "Maternal Smoking During Pregnancy and Smoking by Adolescent Daughters," *American Journal of Public Health* (1994), vol. 84, no. 9.

177 *prenatal exposure to . . . alcohol . . . can change the fetal brain in ways that make the offspring more likely to experiment with or become addicted to such drugs as adolescents and adults*: Rosa Alati et al., "In Utero Alcohol Exposure and Prediction of Alcohol Disorders in Early Adulthood: A Birth Cohort Study," *Archives of General Psychiatry* (2006), vol. 63, no. 9, and John S. Baer and others, "A 21-Year Longitudinal Analysis of the Effects of Prenatal Alcohol Exposure on Young Adult Drinking," *Archives of General Psychiatry* (2003), vol. 60, no. 4.

177 *prenatal exposure to . . . cocaine can change the fetal brain in ways that make the offspring more likely to experiment with or become addicted to such drugs as adolescents and adults*: Lindsey R. Hamilton, H. Donald Gage, and Michael A. Nader, "Altered D2 Receptor Availability in Adult Rhesus Monkeys Exposed to Cocaine In Utero," *The FASEB Journal* (2007), vol. 21, and C.J. Malanga et al., "Augmentation of Cocaine-Sensitized Dopamine

Release in the Nucleus Accumbens of Adult Mice Following Prenatal Cocaine Exposure," *Developmental Neuroscience* (2009), vol. 31.

177 *linked prenatal lead exposure to higher rates of juvenile delinquency and adult crime*: See, e.g., Kim N. Dietrich et al., "Early Exposure to Lead and Juvenile Delinquency," *Neurotoxicology and Teratology* (2001), vol. 23, no. 6; Herbert L. Needleman et al., "Bone Lead Levels in Adjudicated Delinquents: A Case Control Study," *Neurotoxicology and Teratology* (2002), vol. 24, no. 6; and John Paul Wright et al., "Association of Prenatal and Childhood Blood Lead Concentrations with Criminal Arrests in Early Adulthood," *PLoS Medicine* (2008), vol. 5, no. 5.

177 *Sigmund Freud supplied one influential answer*: Sigmund Freud, "Certain Neurotic Mechanisms in Jealousy, Paranoia, and Homosexuality," *International Journal of Psychoanalysis* (1923), vol. 4. See also Peter Gay, *The Tender Passion: The Bourgeois Experience: Victoria to Freud*, Vol. 2 (New York: W. W. Norton & Company, 1999), p. 252.

177 *a specific region of the X chromosome was found more often in gay than in straight men*: Dean H. Hamer et al., "A Linkage Between DNA Markers on the X Chromosome and Male Sexual Orientation," *Science* (1993), vol. 261, no. 5119.

177 *identical twin brothers of gay men are more likely to be gay than are fraternal twin brothers of gay men*: See, e.g., J. Michael Bailey and Richard C. Pillard, "A Genetic Study of Male Sexual Orientation," *Archives of General Psychiatry* (1991), vol. 48, no. 12, and Kenneth S. Kendler et al., "Sexual Orientation in a U.S. National Sample of Twin and Nontwin Sibling Pairs," *American Journal of Psychiatry* (2000), vol. 157, no. 11.

178 *"fraternal birth order effect"*: For a review, see Ray Blanchard, "Fraternal Birth Order and the Maternal Immune Hypothesis of Male Homosexuality," *Hormones and Behavior* (2001), vol. 40, no. 2.

178 *can even be found in the data of the pioneering sex researcher Alfred Kinsey*: Anthony F. Bogaert, Ray Blanchard, and Lesley E. Crosthwait, "Interaction of Birth Order, Handedness, and Sexual Orientation in the Kinsey Interview Data," *Behavioral Neuroscience* (2007), vol. 121, no. 5.

178 *"Observation has directed my attention to several cases"*: Freud, "Certain Neurotic Mechanisms in Jealousy, Paranoia, and Homosexuality."

178 *"The hypothesis doesn't explain all homosexuality, of course"*: Author interview with Anthony Bogaert. See also James M. Cantor et al., "How Many Gay Men Owe Their Sexual Orientation to Fraternal Birth Order?" *Archives of Sexual Behavior*, vol. 31, no. 1.

179 *Bogaert's study of nearly a thousand men*: Anthony F. Bogaert, "Biological Versus Nonbiological Older Brothers and Men's Sexual Orientation," *Proceedings of the National Academy of Sciences* (2006), vol. 103, no. 28.

179 *"a prenatal environment that fosters homosexuality in her younger sons"*: David A. Puts, Cynthia L. Jordan, and S. Marc Breedlove, "O Brother, Where Art Thou?: The Fraternal Birth-Order Effect on Male Sexual Orientation," *Proceedings of the National Academy of Sciences* (2006), vol. 103, no. 28.

180 *in one recent study,* seven times *more likely*: Dana Dabelea et al., "Association of Intrauterine Exposure to Maternal Diabetes and Obesity with Type 2 Diabetes in Youth: The SEARCH Case-Control Study," *Diabetes Care* (2008), vol. 31, no. 7.

180 *a study that has followed a large group of Pima Indians since 1965*: See, e.g., Dana Dabelea, William C. Knowler, and David J. Pettitt, "Effect of Diabetes in Pregnancy on Offspring: Follow-Up Research in the Pima Indians," *Maternal and Fetal Medicine* (2000), vol. 9, no. 1, and Dana Dabelea and David J. Pettitt, "Intrauterine Diabetic Environment Confers Risks for Type 2 Diabetes Mellitus and Obesity in the Offspring, in Addition to Genetic Susceptibility," *Journal of Pediatric Endocrinology and Metabolism* (2001), vol. 14, no. 8.

180 *"a risk of diabetes in the offspring that is over and above any genetic susceptibility"*: Author interview with Dana Dabelea.

181 *"I'm interested in how people think about their diabetes"*: Author interview with Daniel Benyshek.

181 *his studies of the Pima and other Native American tribes*: See, e.g., Daniel C. Benyshek, "Type 2 Diabetes and Fetal Origins: The Promise of Prevention Programs Focusing on Prenatal Health in High Prevalence Native American Communities," *Human Organization* (2005), vol. 64, no. 2, and Daniel C. Benyshek, John F. Martin, and Carol S. Johnston, "A Reconsideration of the Origins of the Type 2 Diabetes Epidemic Among Native Americans and the Implications for Intervention Policy," *Medical Anthropology* (2001), vol. 20, no. 1.

181 *has named this set of attitudes "surrender"*: David L. Kozak, "Surrendering to Diabetes: An Embodied Response to Perceptions of Diabetes and Death in the Gila River Indian Community," *Omega: Journal of Death and Dying* (1996), vol. 35, no. 4.

184 *"When I first read the work of David Barker on birth weight"*: Author interview with Matthew Gillman.

184 *he published an article charting the evolution of his thinking*: Matthew W. Gillman and Janet W. Rich-Edwards, "The Fetal Origin of Adult Disease: From Sceptic to Convert," *Paediatric and Perinatal Epidemiology* (2000), vol. 14, no. 3.

185 *children of women who have a higher intake of Vitamin D during pregnancy*: Carlos A. Camargo et al., "Maternal Intake of Vitamin D During Pregnancy

and Risk of Recurrent Wheeze in Children at 3 Years of Age," *American Journal of Clinical Nutrition* (2007), vol. 85, no. 3.

185 *greater fish consumption during pregnancy was associated with better infant cognition*: Emily Oken et al., "Maternal Fish Consumption, Hair Mercury, and Infant Cognition in a U.S. Cohort," *Environmental Health Perspectives* (2005), vol. 113, no. 10.

186 *Women who gain less than the recommended amount of weight during pregnancy*: Emily Oken et al., "Gestational Weight Gain and Child Adiposity at Age 3 Years," *American Journal of Obstetrics and Gynecology* (2007), vol. 196, no. 4.

186 *Data from Project Viva were instrumental in devising the most recent set of recommendations*: Institute of Medicine, *Weight Gain During Pregnancy: Reexamining the Guidelines* (Washington, DC: The National Academies Press, 2009).

187 *"There does seem to be a lot that a woman can do during pregnancy"*: Author interview with Emily Oken.

190 *an average of two hundred industrial chemicals and pollutants in the cord blood of newborns tested*: Environmental Working Group, "Body Burden: The Pollution in Newborns," July 14, 2005, available at http://reports.ewg.org/reports/bodyburden2/.

190 *The College of American Pathologists now recommends that hospitals store placentas*: Claire Langston et al., "Practice Guideline for Examination of the Placenta: Developed by the Placental Pathology Practice Guideline Development Task Force of the College of American Pathologists," *Archives of Pathology and Laboratory Medicine* (1997), vol. 121, no. 5.

190 *Placentas have even become evidence in the courtroom*: See, e.g., Liz Lempert, "Placentas in the Courtroom," *Technology Review* (1997), vol. 100, no. 1.

191 *"Amniotic fluid provides a window into the child's past"*: Author interview with Simon Baron-Cohen.

191 *He and his team have tested the amniotic fluid of about 200 fetuses*: For a full description of the Cambridge Fetal Testosterone Project, see Simon Baron-Cohen, Svetlana Lutchmaya, and Rebecca Knickmeyer, *Prenatal Testosterone in Mind: Amniotic Fluid Studies* (Cambridge, MA: MIT Press, 2006).

191 *reduced eye contact at one year of age*: Svetlana Lutchmaya, Simon Baron-Cohen, and Peter Raggatt, "Foetal Testosterone and Eye Contact in 12 Month Old Infants," *Infant Behavior and Development* (2002), vol. 25.

191 *more limited vocabulary at two years of age*: Svetlana Lutchmaya, Simon Baron-Cohen, and Peter Raggatt, "Foetal Testosterone and Vocabulary Size in 18- and 24-Month-Old Infants," *Infant Behavior and Development* (2002), vol. 24, no. 4.

191 *more social difficulties at four years of age*: Rebecca Knickmeyer et al., "Foe-

tal Testosterone, Social Relationships, and Restricted Interests in Children," *Journal of Child Psychology and Psychiatry* (2005), vol. 46, no. 2.

191 *greater difficulties with empathy at eight years of age*: Emma Chapman et al., "Fetal Testosterone and Empathy: Evidence from the Empathy Quotient (EQ) and the 'Reading the Mind in the Eyes' Test," *Social Neuroscience* (2006), vol. 1, no. 2.

191 *more narrow interests in childhood*: Knickmeyer, "Foetal Testosterone, Social Relationships, and Restricted Interests in Children."

191 *a stronger interest in systems (such as finding out how things work)*: Bonnie Auyeung et al., "Foetal Testosterone and the Child Systemizing Quotient," *European Journal of Endocrinology* (2006), vol. 155, no. 1.

191 *a greater number of autistic traits*: Bonnie Auyeung et al., "Foetal Testosterone and Autistic Traits," *British Journal of Psychology* (2009), vol. 100.

192 *"I always ask my patients what their birth weight was"*: Author interview with Mary-Elizabeth Patti.

192 *By providing pregnant rats with an antioxidant*: Gilles Cambonie et al., "Antenatal Antioxidant Prevents Adult Hypertension, Vascular Dysfunction, and Microvascular Rarefaction Associated with In Utero Exposure to a Low-Protein Diet," *American Journal of Physiology: Regulatory, Integrative and Comparative Physiology* (2007), vol. 292, no. 3.

192 *By giving pregnant rabbits compounds that block a particular enzyme*: Haitao Ji and others, "Selective Neuronal Nitric Oxide Synthase Inhibitors and the Prevention of Cerebral Palsy," *Annals of Neurology* (2009), vol. 65, no. 2.

192 *By injecting the hormone leptin into newborn rat pups*: Peter D. Gluckman and others, "Metabolic Plasticity During Mammalian Development Is Directionally Dependent on Early Nutritional Status," *Proceedings of the National Academy of Sciences* (2007), vol. 104, no. 31.

192 *By feeding pregnant rats genistein, a compound found in soy*: Dana C. Dolinoy, "Maternal Nutrient Supplementation Counteracts Bisphenol A-Induced DNA Hypomethylation in Early Development," *Proceedings of the National Academies of Science* (2007), vol. 104, no. 32.

192 *by administering the nutrient choline to rat pups prenatally exposed to alcohol*: Jennifer D. Thomas and others, "Choline Supplementation Following Third-Trimester-Equivalent Alcohol Exposure Attenuates Behavioral Alterations in Rats," *Behavioral Neuroscience* (2007), vol. 121, no. 1.

193 *Zinc supplements given to pregnant mice have also been shown to have a protective effect*: Brooke L. Summers, Allan M. Rofe, and Peter Coyle, "Dietary Zinc Supplementation Throughout Pregnancy Protects Against Fetal Dysmorphology and Improves Postnatal Survival After Prenatal Ethanol

Exposure in Mice," *Alcoholism: Clinical and Experimental Research* (2009), vol. 33. no. 4.

193 *consider the case of Charles Gaston*: See, e.g., Janet Golden, *Message in a Bottle: The Making of Fetal Alcohol Syndrome* (Cambridge: Harvard University Press, 2005), p. 159.

193 *"nothing less than child abuse through the umbilical cord"*: Pete Wilson, quoted in Janet Golden, "An Argument That Goes Back to the Womb": The Demedicalization of Fetal Alcohol Syndrome, 1973–1992," *Journal of Social History* (1999), vol. 33, no. 2.

194 *legal scholar Alan Dershowitz has called fetal alcohol syndrome an "abuse excuse"*: Alan M. Dershowitz, *The Abuse Excuse, and Other Cop-Outs, Sob Stories, and Evasions of Responsibility* (Boston: Little Brown, 1994), p. 18.

194 *the dangers of "infant determinism"*: Jerome Kagan, *Three Seductive Ideas* (Cambridge, MA: Harvard University Press, 1998), p. 3.

194 *"Turning to the womb to explain complex social and public-health problems"*: Darshak Sanghavi, "Womb Raider: Do Future Health Problems Begin During Gestation?" *Slate*, October 10, 2008.

195 *"everyone has been changed by their experience in fetal life"*: Author interview with David Barker.

EIGHT MONTHS

199 *"In this world, pregnancy is considered news"*: Joe Dolce, quoted in Rebecca Traister, "Pregnancy Porn: Wacky Names! Baby 'Bumps'! The 'Most Anticipated Baby in the World'!: Why Do We Salivate Over Spawning Celebrities?" *Salon*, July 21, 2004.

199 *actress Demi Moore, who appeared naked and seven months pregnant*: See Annie Leibovitz, "Annie Gets Her Shot," *Vanity Fair*, October 2008.

200 *"compensating for an ungratified curiosity to know where babies come from"*: A. William Liley, "Experiences with Uterine and Fetal Instrumentation," in *Intrauterine Fetal Visualization: A Multidisciplinary Approach*, edited by Michael M. Kaback and Carlo Valenti (New York: Elsevier, 1974), p. 70.

201 *the story of a virus*: See, e.g., John M. Barry, *The Great Influenza: The Epic Story of the Deadliest Plague in History* (New York: Viking, 2004); Alfred W. Crosby, *America's Forgotten Pandemic: The Influenza of 1918* (Cambridge: Cambridge University Press, 1989); and Gina Kolata, *Flu: The Story of the Great Influenza Pandemic of 1918 and the Search for the Virus That Caused It* (New York: Farrar, Straus and Giroux, 1999).

202 *"The 1918 flu pandemic offers an exceptional opportunity"*: Author interview with Douglas Almond.

203 *15 percent less likely to graduate from high school, and 15 percent more likely to be poor; the men earned wages that were 5 to 9 percent lower*: Douglas Almond, "Is the 1918 Influenza Pandemic Over?: Long-Term Effects of In Utero Influenza Exposure in the Post-1940 U.S. Population," *Journal of Political Economy* (2006), vol. 114, no. 4.

203 *20 percent more likely to have heart disease or to be disabled as older adults. Even their height was affected*: Bhashkar Mazumder et al., "Lingering Prenatal Effects of the 1918 Influenza Pandemic on Cardiovascular Disease," *Journal of Developmental Origins of Health and Disease* (2010), vol. 1, no. 1.

204 *schoolchildren with prenatal exposure to fallout from the 1986 Chernobyl nuclear disaster*: Douglas Almond, Lena Edlund, and Marten Palme, "Chernobyl's Subclinical Legacy: Prenatal Exposure to Radioactive Fallout and School Outcomes in Sweden," *Quarterly Journal of Economics* (2009), vol. 124, no. 4.

204 *individuals who were gestated during the period of scarcity that accompanied China's Great Leap Forward*: Douglas Almond et al., "Long-Term Effects of the 1959–1961 China Famine: Mainland China and Hong Kong," National Bureau of Economic Research, Working Paper No. 13384, issued September 2007.

204 *the effects on fetuses of fasting during the Islamic holy month of Ramadan*: Douglas Almond and Bhashkar Mazumder, "Health Capital and the Prenatal Environment: The Effect of Maternal Fasting During Pregnancy," National Bureau of Economic Research, Working Paper No. 14428, issued October 2008.

205 *when pregnancies go well, the benefits are shared by us all—and when prenatal conditions are poor, the price paid by society can be high*: See, e.g., Harold Alderman and E.R. Behrman, "Reducing the Incidence of Low Birth Weight in Low-Income Countries Has Substantial Economic Benefits," *World Bank Research Observer* (2006), vol. 21, no. 1, and Eric I. Knudsen et al., "Economic, Neurobiological, and Behavioral Perspectives on Building America's Future Workforce," *Proceedings of the National Academies of Science* (2006), vol. 103, no. 27.

206 *"the Warren Harding error"*: Malcolm Gladwell, *Blink: The Power of Thinking Without Thinking* (New York: Little Brown, 2005), p. 76.

206 *male CEOs of Fortune 500 companies are three inches taller than the average American man*: Gladwell, *Blink*, p. 87.

206 *each additional inch of height above the average is worth $789 a year in salary*: Gladwell, *Blink*, p. 88.

206 *we are "absurdly biased in favor of the tall"*: Gladwell, *Blink*, p. 96.

206 *A universally shared "unconscious bias" gives tall people an undeserved advantage in hiring and promotion, he writes: "We see a tall person and we swoon"*: Gladwell, *Blink*, p. 88.

207 *"Height is positively associated with cognitive ability, which is rewarded in the labor market"*: Author interview with Anne Case.

207 *examined the height records and test scores of two large groups of Britons and Americans*: Anne Case and Christina Paxson, "Stature and Status: Height, Ability, and Labor Market Outcomes," *Journal of Political Economy* (2008), vol. 116, no. 3.

207 *"Compare a child who has a dozen healthy experiences each day in utero"*: Howard Gardner, "Cracking Open the IQ Box," *The American Prospect*, December 1, 1995.

207 *a landmark attempt to determine the contribution of the fetal environment to IQ*: Bernard Devlin, Michael Daniels, and Kathryn Roeder, "The Heritability of IQ," *Nature* (1997), vol. 388, no. 6641. See also Matthew McGue, "The Democracy of the Genes," *Nature* (1997), vol. 388, no. 6641.

208 *their conclusion that the heritability of IQ is 60 percent*: Richard J. Herrnstein and Charles Murray, *The Bell Curve: Intelligence and Class Structure in American Life* (New York: Free Press, 1994), p.105.

208 *"It made me even more careful"*: Kathryn Roeder, quoted in Byron Spice, "Delivering a Smarter Baby," *Pittsburgh Post-Gazette*, July 31, 1997.

208 *an unusually extensive data set on twins born in Norway*: Sandra E. Black, Paul J. Devereux, and Kjell G. Salvanes, "From the Cradle to the Labor Market?: The Effect of Birth Weight on Adult Outcomes," *Quarterly Journal of Economics* (2007), vol. 122, no. 1.

209 *African-American pregnant women are more likely*: See, e.g., Tyan Parker Dominguez et al., "Racial Differences in Birth Outcomes: The Role of General, Pregnancy, and Racism Stress," *Health Psychology* (2008), vol. 27, no. 2; Cheryl L. Giscombé and Marci Lobel, "Explaining Disproportionately High Rates of Adverse Birth Outcomes Among African Americans: The Impact of Stress, Racism, and Related Factors in Pregnancy," *Psychological Bulletin* (2005), vol. 131, no. 5; and Michael C. Lu and Neal Halfon, "Racial and Ethnic Disparities in Birth Outcomes: A Life-Course Perspective," *Maternal and Child Health Journal* (2003), vol. 7, no. 1.

209 *"Fetal health may be at the fulcrum of black-white disparities in life outcomes"*: Author interview with Douglas Almond.

209 *the introduction of the food stamp program in the late 1960s and early 1970s*: Douglas Almond, Hilary W. Hoynes, and Diane Whitmore Schanzenbach, "Inside the War on Poverty: The Impact of Food Stamps on Birth Outcomes," National Bureau of Economic Research, Working Paper No. 14306, issued September 2008; also forthcoming from *The Review of Economics and Statistics*.

210 *"a future of racial inequality is being programmed"*: Almond, "Is the 1918 Influenza Pandemic Over?"

210 *"merely hatching out embryos: any cow could do that"*: Aldous Huxley, *Brave New World* (New York: Harper Perennial Modern Classics, 2006), p. 8.

211 *"Nothing like oxygen-shortage for keeping an embryo below par"*: Huxley, *Brave New World*, p. 9.

212 *enhanced prenatal and early postnatal conditions account for at least 16 percent of the leap in life expectancy*: Dora L. Costa and Joanna N. Lahey, "Predicting Older Age Mortality Trends," *Journal of the European Economic Association* (2005), vol. 3, no. 2–3.

212 *early attention to maternal health lies behind the famous "French paradox"*: David J. Barker, "Why Heart Disease Mortality Is Low in France: Intrauterine Nutrition May Be Important," *British Medical Journal* (1999), vol. 318, no. 7196.

212 *"It may seem like a cold-blooded thing to say"*: Josephine Baker, quoted in "The World Health Report 2005—Make Every Mother and Child Count," report of the World Health Organization, April 2005, p. 3.

213 *So consistently have military concerns motivated care for pregnant women and infants*: Debórah Dwork, *War Is Good for Babies and Other Young Children: A History of the Infant and Child Welfare Movement in England 1898–1918* (New York: Tavistock, 1987).

213 *"as great as the effects of smoking"*: C. Paul Leeson et al., "Impact of Low Birth Weight and Cardiovascular Risk Factors on Endothelial Function in Early Adult Life," *Circulation* (2001), vol. 103, no. 1264.

214 *"investments targeting fetal health may have higher rates of return"*: Almond, "Is the 1918 Influenza Pandemic Over?"

214 *"early interventions can be much less expensive and much more effective"*: Author interview with Thomas Miller. See also Thomas P. Miller, "Measuring Disparities, Improving Health: Closing the Gap," article produced for the American Enterprise Institute for Public Policy Research, March 17, 2008.

214 *"In contrast to the challenge of changing the determinism of our genes"*: Duane Alexander, statement to the House Subcommittee on Labor-HHS-Education Appropriations, Congressional Hearing on Life Span, April 4, 2001.

215 *exposed a pregnant rat to two commonly used industrial chemicals*: Matthew D. Anway and Michael K. Skinner, "Epigenetic Transgenerational Actions of Endocrine Disruptors," *Endocrinology* (2005), vol. 147, no. 6.

216 *"Most of the patients were exposed from the beginning"*: Leo Kanner, "Problems of Nosology and Psychodynamics of Early Infantile Autism," *American Journal of Orthopsychiatry* (1949), vol. 19, no. 3.

217 *a long and virulent tradition of blaming mothers*: See, e.g., Edward Dolnick, *Madness on the Couch: Blaming the Victim in the Heyday of Psychoanalysis*

(New York: Simon and Schuster, 1998), and *"Bad" Mothers: The Politics of Blame in Twentieth-Century America*, edited by Molly Ladd-Taylor and Lauri Umansky (New York: New York University Press, 1998).

217 *the occasion to coerce or control the behavior of pregnant women*: See, e.g., Cynthia R. Daniels, *At Women's Expense: State Power and the Politics of Fetal Rights* (Cambridge, MA: Harvard University Press, 1993); Laura E. Gomez, *Misconceiving Mothers: Legislators, Prosecutors, and the Politics of Prenatal Drug Exposure* (Philadelphia: Temple University Press, 1997); and Rachel Roth, *Making Women Pay: The Hidden Costs of Fetal Rights* (Ithaca, NY: Cornell University Press, 2000.

217 *a "bio-underclass, a generation of physically damaged cocaine babies"*: Charles Krauthammer, "Children of Cocaine," *Washington Post*, July 30, 1989.

219 *variation in nineteenth-century state laws against abortion*: Elizabeth Oltmans Ananat and Joanna N. Lahey, "The Marginal Child Throughout the Life Cycle: Evidence from Early Law Variation," paper presented at the American Economics Society Meetings, January 3, 2010.

220 *"I can't get my mind off it"*: Author interview with Douglas Almond.

221 *"extracting otherwise inaccessible scientific knowledge from the harsh soil of human catastrophe"*: Richard Neugebauer, "Accumulating Evidence for Prenatal Nutritional Origins of Mental Disorders," *Journal of the American Medical Association* (2005), vol. 294, no. 5.

221 *"The limited role of women's active agency seriously afflicts the lives of all people"*: Amartya Sen, *Development as Freedom* (New York: Knopf, 1999), p. 191.

221 *"hidden penalties of gender inequality"*: Siddiq Osmani and Amartya Sen, "The Hidden Penalties of Gender Inequality: Fetal Origins of Ill-Health," *Economics and Human Biology* (2003), vol. 1, no. 1.

222 *"If I were religious, I would call it divine retribution"*: Amartya Sen, quoted in Shabnam Minwalla, "Hungry Indians Give Sen Food for Thought," *Times of India*, January 10, 2003.

223 *"As I looked around I saw that I was being surrounded by four women"*: Anna Quindlen, "The Ignominy of Being Pregnant in New York City," *The New York Times*, March 27, 1986.

NINE MONTHS

226 *the relationship between Otto Rank and Sigmund Freud*: See, e.g., Samuel Eisenstein, "Otto Rank: The Myth of the Birth of the Hero," in *Psychoanalytic Pioneers*, edited by Franz Alexander, Samuel Eisenstein, and Martin Grotjahn (New York: Transaction, 1995), pp. 36–50; E. James Lieberman, *Acts of Will: The Life and Work of Otto Rank* (New York: Free Press,

1998); and George Makari, *Revolution in Mind: The Creation of Psychoanalysis* (New York: Harper, 2008).

226 *"memory of paradise"*: Otto Rank, quoted in Lieberman, *Acts of Will*, p. 230.

226 *"the unconscious reproduction of the anxiety at birth"*: Otto Rank, *The Trauma of Birth* (New York: Robert Brunner, 1952), p. 12.

226 *The analytic situation re-creates the "intrauterine state"*: Rank, *The Trauma of Birth*, p. 6.

226 *"allowing the patient to repeat with better success"*: Rank, *The Trauma of Birth*, p. 5.

227 *"is prone to be less sensitized"*: Marion Kenworthy, quoted in E. James Lieberman, "Introduction to the Dover Edition," in *The Trauma of Birth* by Otto Rank (New York: Dover, 1994), pp. x–xi.

227 *"a matter of fundamental import"*: Rank, *The Trauma of Birth*, p. 4.

227 *"There is increasing evidence that perinatal stress and pain can have long-term effects"*: Author interview with Vivette Glover.

227 *the level of stress hormones found in babies' umbilical cords*: Rachel Gitau et al., "Umbilical Cortisol Levels as an Indicator of the Fetal Stress Response to Assisted Vaginal Delivery," *European Journal of Obstetrics and Gynecology and Reproductive Biology* (2001), vol. 98, no. 1.

228 *the rise in stress hormones and the intensity of crying in two-month-old infants*: Alyx Taylor, Nicholas M. Fisk, and Vivette Glover, "Mode of Delivery and Subsequent Stress Response," *Lancet* (2000), vol. 355, no 9198.

228 *those who had been circumcised soon after birth reacted more strongly and cried for longer*: Anna Taddio and others, "Effect of Neonatal Circumcision on Pain Responses During Vaccination in Boys," *Lancet* (1995), vol. 345, no. 8945.

228 *"When we do something to a baby that is not an expected part of its normal development"*: Author interview with Anna Taddio.

229 *Early encounters with pain may alter the threshold at which pain is felt later on*: See, e.g., Kanwaljeet J. Anand, "Pain, Plasticity, and Premature Birth: A Prescription for Permanent Suffering?" *Nature Medicine* (2000), vol. 6; Ruth Eckstein Grunau, Liisa Holsti, and Jeroen W. B. Peters, "Long-Term Consequences of Pain in Human Neonates," *Seminars in Fetal and Neonatal Medicine* (2006), vol. 11, no. 4; and Ruth Eckstein Grunau, Michael F. Whitfield, and Julianne H. Petrie, "Pain Sensitivity and Temperament in Extremely Low-Birth-Weight Premature Toddlers and Preterm and Full-Term Controls," *Pain* (1994), vol. 58, no. 3.

229 *Lasting effects might also include emotional and behavioral problems*: See, e.g., American Academy of Pediatrics, "Prevention and Management of Pain in the Neonate: An Update," *Pediatrics* (2006), vol. 118, no. 5.

229 *making their environments quieter, dimmer, and more womb-like*: See, e.g., Heidelise Als and others, "Early Experience Alters Brain Function and Structure," *Pediatrics* (2004), vol. 113, no. 4, and Paul Raeburn, "A Second Womb," *The New York Times Magazine*, August 14, 2005.

229 *Forceps place up to fifty pounds of pressure on a fetus's head, as compared to the nineteen to thirty-three pounds of pressure exerted by the mother's tissues*: Gitau, "Umbilical Cortisol Levels as an Indicator of the Fetal Stress Response to Assisted Vaginal Delivery."

229 *suggest administering pain relief to fetuses undergoing an assisted birth*: See, e.g., Marc Van de Velde et al., "Fetal Pain Perception and Pain Management," *Seminars in Fetal and Neonatal Medicine* (2006), vol. 11, no. 4. See also Annie Murphy Paul, "The First Ache," *The New York Times Magazine*, February 10, 2008.

230 *upwards of 87 percent of pregnant women receive some kind of medication for pain during labor*: Joy L. Hawkins, report presented to the annual meeting of the American Society of Anesthesiologists, October 12, 1999, described here: www.ynhh.org/healthlink/womens/womens_1_00.html.

230 *there were 17.5 percent fewer births on the weekend than expected*: Jeffrey B. Gould, "Neonatal Mortality in Weekend Vs. Weekday Births," *Journal of the American Medical Association* (2003), vol. 289, no. 22.

230 *the average duration of a pregnancy in the United States dropped from forty weeks to thirty-nine weeks*: Michael J. Davidoff et al., "Changes in the Gestational Age Distribution Among U.S. Singleton Births: Impact on Rates of Late Preterm Birth, 1992 to 2002," *Seminars in Perinatology* (2006), vol. 30, no. 1.

232 *soldiers born in spring and summer generally died earlier*: Dora L. Costa and Joanna N. Lahey, "Predicting Older Age Mortality Trends," *Journal of the European Economic Association* (2005), vol. 3, no. 2–3.

232 *recruits with spring or summer birthdays were more likely to develop heart disease*: Dora L. Costa, Lorens A. Helmchen, and Sven Wilson, "Race, Infection, and Arteriosclerosis in the Past," *Proceedings of the National Academies of Science* (2007), vol. 104, no. 33.

232 *70 percent more likely to die of stroke than those born later in the year*: Costa and Lahey, "Predicting Older Age Mortality Trends."

232 *children born in late summer or early fall are a centimeter taller and have thicker bones*: Adrian Sayers and Jonathan H. Tobias, "Estimated Maternal Ultraviolet B Exposure Levels in Pregnancy Influence Skeletal Development of the Child," *Journal of Clinical Endocrinology and Metabolism* (2009), vol. 94, no. 3.

232 *schizophrenics are about 10 percent more likely*: See, e.g., Paolo Castrogiovanni et al., "Season of Birth in Psychiatry: A Review," *Neuropsychobiology*

(1998), vol. 37, no. 4, and E. Fuller Torrey et al., "Seasonality of Births in Schizophrenia and Bipolar Disorder: A Review of the Literature," *Schizophrenia Research* (1997), vol. 28, no. 1.

233 *"I was interested in why individuals are different from one another"*: Author interview with Janet DiPietro.

234 *A recently designed system for assessing the development of the fetus*: Amy Lynn Salisbury, Duncan Fallone, and Barry Lester, "Neurobehavioral Assessment from Fetus to Infant: The NICU Network Neurobehavioral Scale and the Fetal Neurobehavior Coding Scale," *Mental Retardation and Developmental Disabilities Research Reviews* (2005), vol. 11, no. 1.

234 *Richard Bell published a paper containing a very simple observation*: Richard Q. Bell, "A Reinterpretation of the Direction of Effects in Studies of Socialization," *Psychological Review* (1965), vol. 75, no. 2.

234 *Our one-sided focus on the environment the parent provides for the child is "illogical"*: Richard Q. Bell, "This Week's Citation Classic," *Current Contents* (1981), no. 16.

235 *women carrying boys have bigger appetites*: Rulla M. Tamimi et al., "Average Energy Intake Among Pregnant Women Carrying a Boy Compared with a Girl," *British Medical Journal* (2003), vol. 326, no. 7401.

235 *"maternal programming"*: Curt A. Sandman and Laura M. Glynn, "Corticotropin-Releasing Hormone (CRH) Programs the Fetal and Maternal Brain," *Future Neurology* (2009), vol. 4, no. 3.

238 *the fetus moves about once a minute*: Janet A. DiPietro et al., "The Psychophysiology of the Maternal-Fetal Relationship," *Psychophysiology* (2004), vol. 41, no. 4.

238 *The pregnant woman may feel as few as 16 percent of these movements*: Timothy R. Johnson, Elizabeth T. Jordan, and Lisa L. Paine, "Doppler Recordings of Fetal Movement: II. Comparison with Maternal Perception," *Obstetrics and Gynecology* (1990), vol. 76, no. 1.

ACKNOWLEDGMENTS

The trouble with researching and reporting a book while pregnant, I often had occasion to reflect, is that you have a newborn while you're trying to finish it up. For their enthusiastic and unflagging belief in me and this book, I thank my marvelous editor, Dominick Anfuso, and my ace agent, Binky Urban. Leah Miller and Maura O'Brien of Free Press also helped enormously, as did Alison Schwartz and Liz Farrell of ICM. Many thanks, as well, to the Yaddo artists' community for providing an invaluable retreat at an early stage of this project. An admiring thank-you to my editors: Alex Star and Ilena Silverman at *The New York Times Magazine*; Jenny Schuessler at *The New York Times Book Review*; Emily Bazelon, Ann Hulbert, and Hanna Rosin at *Slate*; and Natalie Angier and Jesse Cohen of *The Best Science Writing 2009*. I'm also indebted to the many researchers who shared generously of their time and expertise.

For years now my writers' group has provided me with a near-daily source of support, wisdom, and hard-earned laughter. I offer my thanks to Alissa Quart and Pamela Paul for helping me organize it, and to all the other brilliant and talented members of the Invisible Institute: Gary Bass, Susie Cain, Ada Calhoun,

Elizabeth Devita-Raeburn, Abby Ellin, Sheri Fink, Christine Kenneally, Brendan Koerner, Judith Matloff, Katie Orenstein, Josh Prager, Paul Raeburn, Gretchen Rubin, Kathy Rich, Lauren Sandler, Debbie Siegel, Rebecca Skloot, Stacy Sullivan, Maia Szalavitz, Harriet Washington, and Tom Zoellner. Thanks, too, to the smart, funny, and candid women in my Mums Who Write group.

At crucial moments in the writing of this book, these friends came through: Alison Burns, Susan Burton, Susie Cain, Anna Christensen, Julie Cooper, Elizabeth Devita-Raeburn, Anna Hall, Christine Kenneally, Nadya Labi, Marguerite Lamb, Vanessa Mobley, Ceridwen Morris, Lois Morris, Michele Orecklin, Pamela Paul, Kaja Perina, Alissa Quart, Kathy Rich, Amanda Schaffer, Maia Szalavitz, and Anastasia Toufexis.

Thanks to Colette Stern-Ascher, Andrea Smith, and Aimee Bell for giving me the peace of mind to write. Love and thanks to my mother, father, and sister; my aunt, Roseanne Murphy; and the Witt-Cooper-Staresinic-Maliakal-Symonds clan. And finally, my deepest gratitude to my husband, John, and to my boys, Teddy and Gus: You are my sunshine.

INDEX

INDEX

ABOUT THE AUTHOR

Annie Murphy Paul is a magazine journalist and book author whose writing has appeared in *The New York Times Magazine*, *The New York Times Book Review*, *Slate*, *Discover*, and *The Best American Science Writing*, among other publications. She is also the author of *The Cult of Personality*, a cultural history and scientific critique of personality testing. Read more at www.anniemurphy paul.com or follow her on www.twitter.com/anniemurphypaul.